T0260287

Advancing Engineering Education Beyond COVID

Educators, are you ready to meet the challenge of cultivating the next generation of engineers in a post-COVID-19 context?

Current engineering student cohorts are unique to their predecessors: they are more diverse and have experienced unprecedented disruption to their education due to the COVID-19 pandemic. They will also play a more significant role in contributing to global sustainability efforts. Innovating engineering education is of vital importance for preparing students to confront society's most significant sustainability issues: our future depends on it.

Advancing Engineering Education Beyond COVID: A Guide for Educators offers invaluable insights on topics such as implementing active-learning activities in hybrid modes; developing effective and engaging online resources; creating psychologically safe learning environments that support academic achievement and mental health; and embedding sustainability within engineering education. Students' own perspectives of online learning are also incorporated, with the inclusion of a chapter authored by undergraduate engineering students.

This book consolidates the expertise of leading authorities within engineering education, providing an essential resource for educators responsible for shaping the next generation of engineers in a post-COVID-19 world.

Ivan Gratchev is a Senior Lecturer in Civil Engineering at the School of Engineering and Built Environment, Griffith University, QLD, Australia. He is a Senior Fellow of the Higher Education Academy.

Hugo G. Espinosa is a Senior Lecturer in Electronic Engineering and First Year Coordinator for Engineering at the School of Engineering and Built Environment, Griffith University, QLD, Australia.

Advancing Engineering Education Beyond COVID

A Guide for Educators

Edited by
Ivan Gratchev
Hugo G. Espinosa

CRC Press
Taylor & Francis Group
Boca Raton London New York

CRC Press is an imprint of the
Taylor & Francis Group, an **informa** business

First edition published 2023
by CRC Press
6000 Broken Sound Parkway NW, Suite 300, Boca Raton, FL 33487-2742

and by CRC Press
4 Park Square, Milton Park, Abingdon, Oxon, OX14 4RN

CRC Press is an imprint of Taylor & Francis Group, LLC

ISBN: 978-1-032-20179-5 (hbk)
ISBN: 978-1-032-20312-6 (pbk)
ISBN: 978-1-003-26318-0 (ebk)

DOI: 10.1201/9781003263180

Typeset in Times
by SPi Technologies India Pvt Ltd (Straive)

Contents

THEME 1   COVID-19: Disruption in Context

THEME 2   General Strategies, Approaches, and Practices for Online and Blended Learning Post-COVID-19

THEME 3   Student-Centred Teaching Post-COVID-19: Approaches, Reflections, and Wellbeing

THEME 4   Insights on the Future of Engineering Education

Foreword

Before COVID-19, approaches to engineering education were commonly based on assumptions that students could attend campus for learning activities and assessments. Only a small number of institutions specialised in providing options to undertake much of an engineering programme online.

When governments learned about the risks of COVID-19, engineering educators rapidly converted face-to-face classes to remote teaching of various forms including online classes, simulations, lecture recordings, and laboratories using equipment posted home. As the emergency move to online became longer-term, educators developed capability and learning resources to enhance the remote learning experience for students.

Educators have now invested heavily in remote teaching. Approaches previously achieved by a small number of institutions have become widespread, providing increased flexibility along with challenges. The opportunity now facing engineering educators is to learn from the COVID-19 disruption to engineering education designing and implementing future curricula.

Generic guidance for higher education is often limited in value for engineering educators. The profile of engineering students is different from other disciplines in gender, age, and socio-economic background. Degree programmes involve science, practical laboratories, and engagement with professional practice. Engineering practice is widely diverse in nature.

In *Adopting Post-COVID Teaching Practices in Engineering Education: A Guide for Educators*, Ivan Gratchev and Hugo Espinosa, with the authors of the chapters, provide engineering academics, and professional staff working with engineering students, a comprehensive guide to inform the design of curricula taking advantage of learning from experiences in engineering education since the emergence of COVID-19.

The guide provides international examples of remote learning in engineering; practical chapters with clear, evidence-based principles for teaching remotely; and educators' and students' experiences of remote teaching and learning. Do not miss the essential, evidence-based advice to educators on student wellbeing and stress, and leadership styles that educators can employ to support psychological safety and belonging among students.

Professor Sally Male BE(Hons) PhD FIEAust
Director, Teaching and Learning Laboratory
Faculty of Engineering and Information Technology
The University of Melbourne, Australia

Preface

Ivan Gratchev and Hugo G. Espinosa

COVID-19 has had a profound impact on all aspects of our lives, and teaching is no exception. Due to lockdown restrictions, many universities around the world were forced to rapidly modify their on-campus teaching activities to suit an online learning environment. This period of sudden change and uncertainty produced unprecedented challenges for educators, even those with previous online teaching experience. Given the important role of practical, hands-on activities in developing industry-ready graduates, the transition to purely online teaching was particularly challenging in the engineering discipline. Furthermore, lockdown restrictions and social distancing requirements not only threatened students' continuity of learning but also their physical and mental wellbeing.

Fortunately, technological advances made it possible for many educators and students to continue learning in an online environment during periods of lockdown. Despite the considerable stress created by COVID-19, it also pushed educators out of their comfort zones and forced them to adopt new technologies and teaching approaches that they might not have even considered (or been aware of) before the pandemic. This experience has led to the development and implementation of innovative pedagogical strategies to effectively engage students in online learning.

The changes to teaching enforced by lockdown restrictions may have initially seemed like temporary measures to ensure continuity of learning, which would revert to "normal" once the threat of COVID-19 had passed. However, it is now evident that this period of change has fundamentally shifted educators' and students' views and expectations about education, necessitating the establishment of a "new normal" in a post-COVID-19 learning environment. Moving forward, educators will need to adopt innovative approaches to teaching that are not only safe for students and university staff, but are also effective in engaging students in their learning process and retaining them in their university programme. For engineering educators in particular, the overall impact of the drastic changes arising from COVID-19 have raised important questions, such as:

- How can we effectively engage students in online learning, and provide them with sufficient practical experience needed to become industry-ready graduates?
- How can we integrate online learning activities when variations in access to technology and online learning resources exist?
- How can we design teaching activities that support students' mental health and wellbeing, and align with their expectations and learning styles?

During the process of synthesising research for this book, it was evident that there is no "one size fits all" approach for all subjects; what may be successful for a certain engineering discipline/unit may not work well for others. However, there is still an urgent need to share these learning and teaching practices among academic communities and discuss effective strategies and approaches that can be implemented post-COVID-19.

This book showcases successful learning and teaching practices developed and implemented by leading educators from different countries and universities. It also provides practical advice on how to adopt different blended learning strategies, develop effective online resources to improve student engagement, and modify assessment plans to maintain academic integrity. Some of the book chapters discuss various approaches that can be adopted by a wider community of educators and practitioners, and implemented in courses with different delivery modes, class sizes, and student cohorts. Specific chapters within this book also offer insights on student stress, mental health, and wellbeing, which play a key role in shaping the quality of students' university experience and, ultimately, their academic performance.

This book is written for educators and practitioners across all engineering disciplines, who engage in reflective teaching practices and prioritise student engagement and learning experience. It covers important topics related to learning and teaching strategies that can be adopted effectively post-COVID-19. The book is organised into 12 chapters, grouped according to four key themes.

Theme 1: COVID-19: Disruption in Context (Chapter 1)

Chapter 1 provides an account of the development of engineering disciplines in different countries over the past hundred years, the evolution of student perspectives and industry requirements, and recent changes in learning and teaching methods, including the effect of COVID-19 on student engagement and learning experience.

Theme 2: General Strategies, Approaches, and Practices for Online and Blended Learning Post-COVID-19 (Chapters 2–8)

In Chapter 2, the authors provide recommendations on how to adopt a blended learning environment and describe various techniques on how to combine online activities with face-to-face sessions. This chapter also provides strategies that could assist educators to better engage in blended learning post-COVID-19.

Chapter 3 describes practical insights into the development of online resources and provides guidelines on how to create successful online videos that will help to engage students in the learning process.

Chapter 4 discusses the experience with teaching large classes of science students during the pandemic, when advanced online tools such as learning journals combined with personal student plans were implemented by the teaching team to engage students remotely and regularly monitor their activities.

Chapter 5 discusses the student and academic experience with conducting laboratory work online due to the COVID-19 restrictions. A set of pre-recorded videos of experimental procedures, online sessions with a student-teacher interaction, and follow-up assessment quizzes were introduced in two engineering mechanics courses.

Chapter 6 highlights the changes brought about by COVID-19 and identifies the need for new skills required by industry. Living with the virus requires relevant changes in academic curricula with a focus on the development of new programmes and more rigid evaluation systems.

Chapter 7 discusses the implementation of an online course on Microwave Engineering containing a comprehensive hands-on remote laboratory component. The motivation, structure, and outcomes of each lab are discussed, as well as student feedback based on their learning experience.

Chapter 8 describes how engineering teaching staff from two academic units within the University of Auckland have adapted to remote, "emergency" teaching. Staff have developed flexible and innovative approaches to offer both theory and practical components to students. Two undergraduate geotechnical engineering courses are discussed.

Theme 3: Student-Centred Teaching Post-COVID-19: Approaches, Reflections, and Wellbeing (Chapters 9–11)

Chapter 9 provides an overview of the stress process as a factor that contributes to student retention and academic success. The pressure experienced by students has increased due to COVID-19, and this chapter discusses the major principles and strategies that can be effectively employed by educators to enhance students' capacity to cope with stress.

Chapter 10 presents and discusses the results of student surveys, which identify students' uncertainties and common challenges that they have faced and dealt with throughout the pandemic. This chapter also suggests strategies that can be readily implemented to improve students' learning experiences.

Led by undergraduate students, Chapter 11 reveals and discusses student perspectives and reflections about online teaching that may be unknown to educators. This chapter not only shares students' experiences, but it also recommends valuable ideas and strategies to educators on how to develop engaging courses.

Theme 4: Insights on the Future of Engineering Education (Chapter 12)

Chapter 12 concludes the book. It describes the challenges faced by universities, educators, and students post-COVID-19, as well as some suggestions on how to approach them. In addition, the chapter outlines the approaches that educators may adopt in the near future when the pandemic becomes endemic. Finally, the chapter explores one of the most significant challenges facing engineering education: equipping our future engineers with the knowledge and skills to contribute to global efforts to achieve the United Nation's Sustainable Development Goals by 2030.

Acknowledgements

The editors would like to thank all authors of this book for sharing their knowledge, expertise, and inspiring stories; this book would not exist without their valuable contributions.

We would like to thank the School of Engineering and Built Environment at Griffith University, Australia, for their commitment to championing engaging engineering education, and for supporting the development of this book through the Scholarship Activities Learning & Teaching Grant 2021.

We are deeply grateful to Mrs Janelle Cruickshank (www.theproofleader.com.au; admin@theproofleader.com.au) for her thorough and extraordinary job in copyediting every chapter of this book.

Special thanks go to Dr Amanda Biggs for her invaluable support, editing assistance, and suggestions for shaping the structure and style of this book.

We are profoundly thankful to our families for their constant support and patience during the arduous process of writing this book.

Finally, our sincere thanks to Mr Marc Gutierrez and Ms Kianna Delly from CRC Press for their continuous guidance and support in making the idea of this book a reality.

Notes on the Editors

Dr Ivan Gratchev is a Senior Lecturer at the School of Engineering and Built Environment, Griffith University, QLD, Australia. He graduated from Moscow State University (Russia) and received his PhD from Kyoto University (Japan). He worked as a research fellow in the geotechnical laboratory of the University of Tokyo (Japan) before joining Griffith University (Australia) as a lecturer. His research interests are in geotechnical aspects of landslides, soil liquefaction, and rock mechanics. He has published numerous research articles in leading international journals and international conferences.

Since joining Griffith University in 2010, Dr Gratchev has taught several engineering courses (including soil mechanics, rock mechanics, and geotechnical engineering practice) using a project-based approach. His teaching achievements were recognised by his peers and students through several learning and teaching citations and awards, including Senior Fellow of the Higher Education Academy (SFHEA). Dr Gratchev's innovative approach to teaching resulted in publication of two textbooks which have become a popular study resource among students around the world. Dr Gratchev is the Chair of Webinars in Engineering Education, which provide a platform for academics and practitioners to share their learning and teaching practices.

Dr Hugo G. Espinosa is a Senior Lecturer in electronic engineering at the School of Engineering and Built Environment, Griffith University, QLD, Australia. He is the First Year Coordinator for Engineering (Nathan Campus), and the Program Director for the Diploma of Engineering. Dr Espinosa received his bachelor's degree in Electronic and Telecommunications Engineering from the Monterrey Institute of Technology and Higher Education, Mexico; his master's degree from the University of Sao Paulo, Sao Paulo, Brazil; and his Ph.D. degree (Summa Cum Laude) from the Technical University of Catalonia, Barcelona, Spain, both in Electronic Engineering. He has been a visiting researcher at the Federal Polytechnic School of Lausanne, Lausanne, Switzerland, and a Postdoctoral Fellow at the School of Electrical Engineering, Tel Aviv University, Tel Aviv, Israel. Dr Espinosa is the Chair of the Antennas and Propagation Chapter of IEEE QLD Section. He is a Member of the IEEE Education Society, the IEEE Antennas and Propagation Education Committee, and the Australasian Association for Engineering Education. He is also an IEEE STEM Ambassador. His approach to teaching is based on innovative techniques such as experiential learning, project based-learning, and flipped classroom; his evidence-based teaching methods have been published in journal articles, conference proceedings, and book chapters. Dr Espinosa has received the Griffith Engineering Lecturer of the Year award in 2020 and 2021, and the Senior Deputy Vice-Chancellor Excellence in Teaching commendation in 2015 and 2019, and the IEEE QLD Section Outstanding Volunteer award in 2022.

Notes on the Contributors

Sanam Aghdamy is an experienced engineer with a grounding in many fields of structural engineering such as sensor technology, impact/blast response of structures and composite structures before establishing a specialized career in facade engineering. She has demonstrated history of working in academia (research and teaching) for over 10 years and is an associate Fellow of the UK Higher Education Academy.

Amanda Biggs is a lecturer in the Department of Employment Relations and Human Resources at Griffith University, Australia. She convenes undergraduate and post-graduate courses covering topics including stress, mental health, leadership, and work-life balance. Her research on stress, burnout, and work engagement has been published in leading international journals.

Martin S. Brook is an Associate Professor at Auckland University, New Zealand. His research interests include engineering geology and education, with a focus on developing work integrated learning opportunities for students.

Andrew Busch is the Deputy Head of School Learning and Teaching at the School of Engineering and Built Environment, Griffith University, Australia. He has been Director of Work Integrated Learning for the Sciences Group at Griffith University. His research in engineering education includes experiential learning and project-based learning.

Savindi Caldera is a Research Fellow at the Cities, Research Institute, Griffith University, Australia. Her research in engineering education focuses on sustainability competencies and problem-based learning. A recipient of several prestigious early career academic awards, including the Griffith Highly Commended Award for Excellence in Teaching and QUT Vice-Chancellor's Performance award. She is a Fellow of the UK Higher Education Academy.

Patricia Caratozzolo is a Full Researcher at the Institute for the future of Education, and Associate Professor at the School of Engineering and Sciences, Tecnológico de Monterrey, Mexico. Her research includes educational innovation, critical thinking, skills for Industry 4.0, and social-oriented interdisciplinary projects in STEAM.

Julie Crough is a Grants Coordinator at Gilmour Space Technologies, and Adjunct Research Fellow at Griffith University, Australia. Dr Crough has been a Learning and Teaching Consultant at Griffith University. She is the co-editor of the book *Blended Learning Designs in STEM Higher Education* published by Springer.

Sam Cunningham is a Lecturer in Learning and Teaching Development, Impact and Recognition at Queensland University of Technology, Australia. His research areas include engineering education and natural language processing. He is a Senior Fellow and Associate Fellow (Indigenous) of the Higher Education Academy.

Sarah Dart is Strategic Lead for Learning and Teaching Development, Impact and Recognition at Queensland University of Technology, Australia. Her research includes the impact of worked example videos as part of blended learning designs. Dr Dart has received numerous awards, including the Early Career Teaching Excellence Award from the Australian Mathematical Society in 2021.

Cheryl Desha is a Professor at the School of Engineering and Built Environment, Griffith University, Australia. Professor Desha was awarded Engineers Australia's Queensland's Professional Engineer of the Year in 2021, and in 2020 was awarded the Queensland Government's Individual Champion of Change Award by the Inspector General Emergency Management.

Peter Doe is Honorary Research Associate at the University of Tasmania, Australia. He has written numerous book chapters, and has published more than 80 journal and conference papers, including five on engineering education.

Hugo G. Espinosa is a Senior Lecturer in Electronic Engineering at the School of Engineering and Built Environment, Griffith University, Australia. He is the First-Year Coordinator for Engineering (Nathan Campus). His research in engineering education includes experiential learning, project-based learning, and flipped classroom. He received the Griffith Engineering Lecturer of the Year award in 2020 and 2021.

Cynthia Furse is Professor in the Electrical and Computer Engineering Department at the University of Utah, USA. Dr. Furse is a Fellow of the IEEE and the National Academy of Inventors. She is a leader in the flipped-classroom teaching method and has received numerous teaching and research awards including the 2020 IEEE Chen-To Tai Distinguished Educator Award.

Ivan Gratchev is a Senior Lecturer at the School of Engineering and Built Environment, Griffith University, Australia. Dr. Gratchev is the author and co-author of the books *Soil Mechanics through Project-Based Learning*, and *Rock Mechanics through Project-Based Learning*, both published by CRC Press. He is a Senior Fellow of the Higher Education Academy.

Alexander Gregg is a lecturer within the Discipline of Aerospace, Mechanical and Mechatronics Engineering at the University of Newcastle, Australia. His research in engineering education includes factors that influence student engagement with educational video content. He was the recipient of the 2018 Vice Chancellor's Award for Teaching Excellence and Contribution to Student Learning.

Lillian Guan is an undergraduate student in electronics engineering and physics at Griffith University, Australia. She has been actively involved with Women in Engineering, and Ladies in Technology, Engineering and Science at Griffith University. She serves as a tutor and Peer-Assisted Study Session (PASS) program leader.

Shanmuganathan Gunalan is a Senior Lecturer at the School of Engineering and Built Environment, Griffith University, Australia. He completed the Graduate Certificate in Higher Education, and has been the recipient of Griffith University Grants for Learning and Teaching.

Alan Henderson is an Associate Professor at the School of Engineering, University of Tasmania, Australia. He is an enthusiastic educator with interests in the scholarship of university learning and teaching. He has worked as an investigator of several office of learning and teaching national projects.

Damien Holloway is an Associate Professor and Associate Head of Learning and Teaching in Engineering at School of Engineering, University of Tasmania, Australia. Professor Holloway's teaching has been predominantly in the areas of structural mechanics, stress analysis and vibrations. He has authored and co-authored over 130 refereed journal and conference papers.

Maria Kovaleva is a Lecturer at Curtin University, Australia, and a Fulbright Scholar at Brigham Young University, USA. Dr Kovaleva has been teaching electromagnetics and antennas since 2018. Her research at Curtin Institute of Radio Astronomy includes electromagnetic analysis of SKA-Low.

Vianney Lara-Prieto is the Division Director of the School of Engineering and Sciences at Tecnológico de Monterrey, and Adjoint Researcher at the Institute for the future of Education. Her research includes challenge-based learning, social-oriented education, educational innovation, interdisciplinary STEM education, and women in STEM.

Christopher Love is an Associate Professor and the Deputy Dean Learning and Teaching at Griffith University, Australia. Associate Professor Love is a Senior Fellow of the Higher Education Academy. He has been the recipient of several teaching awards and citations including an Australian Award for University Teaching in 2019 and Griffith Award for Teaching Excellence in 2016 for his innovative teaching practices.

Sarah Lyden is a Senior Lecturer in the School of Engineering, at the University of Tasmania, Australia. Dr Lyden's main areas of research are renewable energies, and the integration of renewable energy sources into the electricity grid. Dr Lyden is actively involved in STEM education and outreach research projects.

Shweta Mehta is an engineering graduate who majored in software engineering. She has been President of the Women in Engineering leadership program, and Vice-President of Engineering for the Ladies in Technology Engineering and Science society, both at Griffith University, Australia.

Jorge Membrillo-Hernández is an Assistant Professor at the Bioengineering Department of the School of Engineering and Sciences at Tecnológico de Monterrey, and Full Researcher at the Institute for the Future of Education. His research includes challenge-based learning, social-oriented education, educational innovation for sustainability, interdisciplinary STEM education, and women in STEM.

Benjamin Millar is a Lecturer in Electrical and Electronics Engineering at the University of Tasmania, Australia. His fields of research include distributed systems and algorithms, electrical energy generation, and signal processing.

Mahan Mohammadi is a PhD Candidate in Healthcare Service Management at Griffith University, Australia.

Hayden Ness is an undergraduate student in electronics and computer science at Griffith University, Australia. Hayden has been executive and president of the Griffith Electronics Engineering Club, and Peer-Assisted Study Session (PASS) program leader.

Rolando P. Orense is a Professor at Auckland University, New Zealand. His research interests include civil geotechnical engineering and engineering education.

Nick P. Richards is a Professional Teaching Fellow at Auckland University, New Zealand. He teaches Earth Sciences courses at both undergraduate and postgraduate levels. Dr Richards is the undergraduate advisor for the Earth Science Programme in the School of Environment.

Belinda Schwerin is the Head of Electronic Engineering at the School of Engineering and Built Environment, Griffith University, Australia. She is an Executive Member of the Engineering and Built Environment Learning and Teaching Committee, at Griffith University. Her research in engineering education includes flipped-classrooms for large classes in advanced mathematics.

Berardi Sensale-Rodriguez is an Associate Professor in the Electrical and Computer Engineering Department at the University of Utah, USA. He has received several awards including the 2019 ECE department Outstanding Teaching award. His research interests include simulation and design of electronic and photonic devices and structures.

Amin Vakili holds a Fellowship of the Royal Australian and New Zealand College of Psychiatrists. He has worked as a General Adult psychiatrist in both public and private sectors, and has particular interest in mood, anxiety, eating, and personality disorders, as well as ADHD, stress and burnout, PTSD, and OCD.

Patricia Vázquez-Villegas is a Doctorate Professional in Biotechnology, and researcher at the Institute for the Future of Education, Socially Oriented Interdisciplinary STEAM (SOI-STEAM) Education Research Group, at Tecnológico de Monterrey, Mexico.

Anna Wrobel-Tobiszewska is a lecturer in Environmental Engineering at the University of Tasmania, Australia. Dr Wrobel-Tobiszewska has expertise in a range of civil and environmental subjects, including engineering curriculum and pedagogy.

Theme 1

COVID-19
Disruption in Context

1 Changes in Student Demographics, Behaviour, and Expectations in Higher Education

Andrew Busch and Belinda Schwerin
Griffith University, Nathan, Australia

CONTENTS

INTRODUCTION

The COVID-19 pandemic has caused considerable disruption to the higher education industry, with a rapid transition to online learning and considerable changes in how students are taught. However, while the pace of this particular transition has been rapid, the last century had already seen significant changes in higher education (Gopalan, 2016). The true impact of the current pandemic can thus only be properly assessed relative to these background trends in student demographics,

DOI: 10.1201/9781003263180-2

learning preferences, and campus experiences. Assessing the impact of COVID-19 also requires a clear understanding of student motivation and expectations, as well as the learning preferences of both students and faculty. Given the important role that intrinsic motivation plays in both driving student behaviour and in student success (Howard et al., 2021), an understanding of what motivates students, and how these factors changed leading into the pandemic, is critical in assessing its ongoing educational impact. As such, this chapter will investigate underlying changes in student demographics, the student experience, and student expectations leading up to and during the COVID-19 pandemic, establishing a valid context in which the recent work in assessing the impact of the pandemic can be viewed.

Accurately analysing trends in higher education is difficult, primarily due to the relative lack of well-controlled longitudinal studies in the field. Most of the research focuses on small cohorts of students from either one institution or a small group, generally confined to a single country, and is often carried out entirely over a single intake. While this gives some local information on short-term trends, long-term trends are less reliable, and show considerable bias and uncertainty when extrapolated to the national or international levels. As such, the approach taken in this work is to analyse a representative sample of the research in these areas, and attempt to draw qualitative conclusions from these individual studies. While no formal meta-analysis is performed, this approach will allow for broad trends and preferences to be identified and examined without relying on quantitative analysis or statistical validation.

CHANGING STUDENT DEMOGRAPHICS IN HIGHER EDUCATION

Over recent decades, there has been much effort to increase the diversity of students attending higher education (HE), partly due to demand. Efforts have differed across countries, but overall, the HE cohort demographics are following similar trends with slowly increasing participation by minority and non-traditional groups, including those from low socio-economic backgrounds, first-in-family, and mature students. However, these trends have not necessarily been reflected in each discipline, and in particular, engineering still struggles to attract student diversity representative of the population, and in sufficient numbers to support industry requirements.

Pre-World War 2

Classical higher education, as mostly seen in the United Kingdom (UK), United States (US), and Australia (AU), was founded on the traditions of previous centuries, consisting of a routine of lectures and recitation in each class. Students typically studied classical languages, maths, and philosophy in their first year, and there were no choices of majors or electives, with success based on effort rather than interest (Snyder, 1993). Classes required students to orally recite content from memory in front of the class, and thus university study was focused on the repetition of the knowledge of instructors, rather than independent, creative, or novel thinking. What's more, students attending university or college were typically only from well-off families, and children of professionals.

The Australian higher education system pre-World War 2 (WW2) was largely modelled on the traditional British university system, and later the "Land Grant" colleges in America, and controlled by the state governments. Only around 0.1% of the population in 1914 were enrolled in university and 0.2% by 1939 (Breen, 2002) and, similarly, these were from professional or well-off families. While the state provided some funding, students typically paid fees to attend university, limiting access to those able to afford it.

Post-World War 2

Post-WW2, there was significant growth in the number of students aged 18–21 attending university or college. Consequently, there was a change in the university climate. In the UK and USA, more public universities opened, compared to the previous climate where universities were largely private and prestigious. The USA and many other countries saw improved affluence, increasing the demand for secondary and higher education. As a result, the number of universities in many countries doubled or tripled, and the push for eradicating illiteracy and providing quality education was a focus (Goldin & Katz, 1999). Industries emerging in this boom time required more modern scientific technology and, consequently, there was a need for the introduction of new methods, processes, and machines. Old skills were becoming redundant, requiring workers to reskill, and new techniques emerged. Scientific knowledge was growing, resulting in the demand for researchers, skilled workers, and high-level professionals (Goldin & Katz, 1999). This all created a demand for additional university places and graduates.

Following WW2, many governments offered schemes to support returning soldiers to go to university. Over the following decade, the number of students aged 17–30 years old attending university in the UK increased to 5% in 1960, 8.4% in 1970, and continued to rise to 47% in 2010 (Douglass, 2012). Yet despite this increase in participation in higher education, little headway had been made in reducing the socio-economic inequality in HE participation (pre-1990) (Crawford, 2012).

In Australia, university enrolments also saw a surge, with ex-servicemen being provided with government-sponsored university places. The federal government also increased its role in financing higher education, as compared to the state governments. Student enrolments increased to 148,000 students by 1975 (K12 Academics, 2004). However, socio-economic inequality was a considerable problem, as university places were funded either through Commonwealth scholarships based on merit, or through fees, which could only be afforded by those from more affluent families. Recognising the need to make universities more accessible to working and middle-class groups, university fees were abolished in the 1970s, increasing the enrolment rate. The Commonwealth government fully funded higher education until the recession and increasing costs of higher student enrolment rates made this unsustainable, and fees were reintroduced in the 1980s.

Despite post-war efforts to increase participation rates, the late 1970s to early 1990s still saw a wide (in some cases widening) gap in the higher education participation rates between rich and poor. Chowdry et al. (2013) compared participation rates of UK youth based on socio-economic status, and found that those in the lowest fifth socio-economic status group had a participation rate 40% lower than those from the highest fifth group. Numerous studies (Crawford, 2014; Devlin, 2013) have also identified that students coming from families where parents work in occupations requiring higher education are less likely to drop out than those from lower social class backgrounds. Consequently, parental income was found to be a key determinant of whether youth would continue on to higher education (Blanden et al., 2002).

Funding for university places underwent changes during this period to ensure they were sustainable. In the mid-1980s, Australia was no longer able to fund the growing number of students entering university and reintroduced student fees. This was later changed to a HECS (Higher Education Contribution Scheme) system in which fees could be deferred for payment after degree completion once income increased above a threshold. Similarly in the UK, means-tested tuition fees were also introduced in 1998, later changing to a deferred payment scheme. HE participation rates in Australia continued to increase from 20% in 1980 to 28% in 1989 and 38% by 1994 (Marks et al., 2000). Studies on participation rates have not seen any significant reduction in HE participation rates by low socio-economic students as a result of this introduction, but there was only a small reduction in the gap between high and low socio-economic participation in HE.

Recent Decades (2000+)

Since 2000, participation rates in HE have continued to rise, and there has been reported to be a narrowing of the gap between low and high Socioeconomic Status (SES) groups over time. A study evaluating students entering university (18–19 years old) in the UK showed a rise in rates from 29.7% in 2004–2005 to 34.4% in 2009–2010 (Crawford, 2014), and 47% for the 17–30-year-old age group (Douglass, 2012). It also showed that there was a larger increase for those from lower SES backgrounds.

Further, recent statistics from the UK show an increase in enrolments for those aged 21–24 and 25–29, but a decrease in students over 30 years old. The increase seen in the 21–29-year-old group has been consistent over the five years leading up to 2019 (prior to the pandemic). By 2019, students aged 21 or older (mature students) at the start of their program made up 39% of undergraduates enrolling in UK universities, 28% of whom studied part-time (Hubble & Bolton, 2021). Students with a declared disability also increased each year. Diversity of ethnicity was also shown to increase each year over that five-year period (Mantle, 2021). While gaps in participation between low and high socio-economic groups have been reduced, they were also found to vary considerably in different regions. The gap narrowing has, in part, contributed to the shift in employment rates for students while studying. While students from the early 2000s undertook mostly seasonal work, such as over holiday periods, closer to 2020 students are more often undertaking ongoing part-time jobs (Office for National Statistics, 2016). This trend is seen in other countries also.

Compared to the US, the UK has lower entry rates to HE, but higher completion rates (lower drop-out rates). In 2013, 64% of UK pupils were identified as likely to enter HE at some time in their life, versus 72% of US pupils, while 80% of UK students complete their degree, compared with under 65% in US (Crawford, 2014).

The link between higher entry rates to HE and drop-out rates has been well established by numerous studies, particularly where additional students enrolling in HE are the first in the family, or where additional student enrolments include those with lower literacy and numeracy skills.

In the US, around 33% of 18–24-year-olds enrolled in college. In particular, there is considerable growth in female student enrolments, with typically more female students than male now enrolled in HE (Snyder, 1993). There is also increased ethnic diversity, with students from vastly different cultural, as well as socio-economic, backgrounds. Students of colour now comprise more than 45% of graduates, compared with 30% in 1996. One-third of students are the first-in-family to attend HE. Clearly, there is a shift in the student populations across HE institutions, with more adult and non-traditional students enrolling, and consequently increased demand for greater flexibility in their education models. There is an increase in underrepresented students, including mature students, those with dependants, and those studying part-time to balance work. These groups have an increased risk of dropping out: Thirty-three per cent of first-in-family drop out within three years compared to 14% of those with parents with a HE degree (Hanover Research, 2020). The cause of higher drop-out rates for first-in-family students has been investigated by studies such as Crawford (2014) and Devlin (2013), which showed that first-in-family students were less familiar with the nuances of university culture and had unrealistic expectations of the time and effort required for success. These students require additional or different support services to help them get through to graduation.

Australia has, similarly, taken steps to improve the inequality in HE cohort demographics. In 2008, the Australian Government lifted restrictions on university enrolments in order to increase the number of graduates and to make universities more accessible to students from lower socio-economic groups. This was known as "demand-led funding", and led to a surge in domestic undergraduate enrolments from 53% to 60% of school leavers (Productivity Commission, 2019). A significant number of additional students, including those from low socio-economic backgrounds, enrolled in HE where they otherwise would not have. However, it did little to improve the participation of rural or Indigenous groups (Productivity Commission, 2019).

The initiative resulted in students of lower ATAR (Australian Tertiary Admissions Rank) (typically <70) enrolling in HE than was previously the case. Low ATAR scores can be a barrier for those who have experienced hardship or who are from disadvantaged backgrounds, restricting their entry into university when they may otherwise show academic promise (Cardak & Ryan, 2009). While there is considerable variance around the outcomes of low ATAR students, on average, high ATAR students achieve higher academic outcomes and lower drop-out rates compared to lower ATAR students. Students with lower ATAR more often have lower literacy and numeracy skills, higher drop-out rates, and longer times to completion. Of the additional students who enrolled in HE during the demand-led funding initiative, who typically underperformed academically relative to other students, and had much higher drop-out rates, 21% dropped out prior to completion versus 12% for the rest of the cohort, and about 25% took much longer to finish (Productivity Commission, 2019). During this same period (2008–2017), direct entry students (generally by alternative pathways rather than straight from grade 12) more than doubled, while admissions via the Tertiary Admissions Centre (TAC) grew by only 17%.

In 2017, the Australian Government ended the demand-led funding structure, with the aim of incentivising universities to provide quality education, rather than just quantity. This meant that admitting additional students did not provide the university with any additional funding.

Current trends in HE participation rates show they increased to where around one in two school leavers will enrol in HE. While the number of low ATAR students enrolling is increasing, this is still a small percentage of students, but there is also a large number of students entering HE via non-traditional pathways such as vocational qualifications. Australia has always had a considerable number of off-campus enrolments, but off-campus enrolments has increased more than on-campus enrolments. Around one-third of students enrol in either multi-modal or off-campus courses in 2016, even before the pandemic, and around 45% of on-campus students study half or more of their courses online (Norton et al., 2018). This has been beneficial to those students with work or family responsibilities, and reduces travel expenses for students.

Parental occupation is still a strong factor for participation, with those whose parents attended university more likely to go themselves, but gaps in HE engagement have reduced across all SES groups. Highest rates of participation in HE occur where parents speak an Asian language in the home, while the lowest rates are found where parents speak English or are from Pacific Islands (Marks et al., 2000). Women now account for more the half of all enrolments in HE, but male–female enrolment ratios vary considerably by discipline. Finally, more than 90% of those applying are offered a place in HE, and students are more inclined to move between courses and universities, using previous study to gain admission (Norton et al., 2018).

ENGINEERING

While university enrolments in general look to have an upward trend, in the engineering field, commencements have been static or falling since the peak in engineering student cohorts of 2015 (Kaspura, 2019). The diversity of students engaging in HE overall has greatly increased, but that in engineering is considerably less representative of the population, favouring Asian, mixed, and Caucasian groups (National Science Board, 2019). While female student numbers are gradually increasing, they are still very underrepresented in engineering and IT programs, and in 2017 accounted for only around 16% of the student cohort (Kaspura, 2019).

In spite of the recent decline in engineering enrolment, many of the trends associated with higher HE participation rates equally apply to engineering, including increases in lower ATAR and first-in-family students. A study of first-year engineering student responses to the challenges of university indicated trends in current student experiences (Devine, 2013). Thirty-seven per cent of full-time students in the study spent more than 15 hours per week in paid work, and 20% worked more than 20 hours each week (Devine, 2013), adding to the challenge of balancing studies with work requirements. This is consistent with finding by the Office for National Statistics (2016) that students engaged more in part-time work throughout their studies, as opposed to previous patterns of holiday work. Other findings included a difference in the way first-generation students handled difficulties with their studies, reasonability of time management and planning, unfamiliarity with expectations and demands of HE study, and difficulties with language, all of which increased the difficulty of the transition to university, and increased the risk of drop-out. Findings of this study support the need for universities to provide additional support so that students have the opportunity to succeed.

CHANGES IN STUDENT ATTENDANCE RATES AND THE ON-CAMPUS EXPERIENCE

Campus attendance and class participation have been shown to be strongly correlated with student success and grades (Bronkema & Bowman, 2019; Credé et al., 2010). The reasons for this correlation are complex, and causation is not proven; however, it seems likely that there is at least some causative factor linking attendance and success. Given its place as a strong indicator of student success, it is critically important to understand the reasons for non-attendance and investigate the mechanisms behind the falling rates of classroom participation, and in this section, we provide an overview of the literature surrounding this topic.

Higher education campus and class attendance has received some attention in the literature; however, there have been very few longitudinal studies investigating trends in campus and class attendance. It is a common trope across higher education that attendance rates at lectures and other classes are falling, but this is not necessarily well supported by evidence, with writings from as early as the 14th century bemoaning the "falling attendance rates" at Oxford University (Tuchman, 2011). Since that time, the general consensus from each generation of academia is that class attendance is "falling sharply"; however, if this were indeed the case, we would be seeing no attendance at all in modern classrooms. On the contrary, attendance in many university programmes is still relatively high, particularly when considering that many students must now juggle employment, family responsibilities, and travel requirements, issues that were previously of lesser impact due to student demographics. Prior to 1970, students in higher education were overwhelmingly male, of high socio-economic background, and unmarried. This is certainly no longer the case, and this higher diversity within the student population means that more students have many competing demands on their time that previously did not exist.

One of the first studies of absenteeism in a higher education setting was conducted by Romer (1993). This study looked at the rates of absenteeism across three universities, with averages of 34%, 39.7%, and 24.8%. Absentee rates were found to be higher in larger courses when compared to smaller courses, and higher in those with less mathematical content. Courses studying more general content, typically in earlier years, also had higher rates on non-attendance when compared to those with more specialised content in later years of study. Another interesting finding from this research was that absenteeism was lower when the courses were taught by regular faculty (34%) compared to those taught by other instructors such as teaching assistants or part-time staff (47%). This is thought to relate to students' perception of teaching quality. The relationship between attendance rate and course results showed a high positive correlation in this study, with attendance accounting for 31% of the variation between grades. This relationship held true even when applying factors to account for student interest in the topic, and prior academic performance. Overall, this study showed a very large and potentially causal relationship between class attendance and student performance.

A large longitudinal study conducted over a 5-year period from 1994 to 1999 also found some large changes around the student experience that occurred in the 1990s (Krause et al., 2005; McInnis et al., 2000). This work found that in 1994, more than 84% of students spent four or more days on campus per week, with an average of 17.6 contact hours. Just five years later, this situation had changed considerably, with the percentage of students spending four or more days on campus falling to 71%, with 17.1 contact hours. In 2004, that figure had increased back to 77%; however, average contact hours had fallen again to 16, with speculation that students were instead spending time on campus using computing facilities that were previously unavailable. Interaction between students also appears to have changed markedly over this same period, with an increase in students reporting that they "never" or "hardly ever" worked with other students in areas of study where they were experiencing problems increasing from 30% in 1994 to 50% in 1999, and then further increasing to 60% in 2004. This change in both on-campus engagement and collaboration with other students appears to be somewhat linked to employment, as the percentage of students engaged in part-time work while studying increased from 43% to 51% from 1994 to 1999, respectively, while remaining constant at 51% in 2004. The number of hours spent in such employment also increased, with the percentage of students working more than 10 hours per week rising from 40% in 1994 to 52% in 1999, then falling slightly in 2004. The authors of this study proposed that the accessibility of online study materials also contributed to lower student collaboration, as students could more easily find solutions to troublesome areas themselves. A summary of the major findings of this work are presented in Table 1.1.

IMPACT OF ATTENDANCE ON STUDENT PERFORMANCE

It has long been suspected that class attendance has a direct and positive impact on student results, and there has been a relatively large amount of research conducted in this area. The following section will give a summary of a representative sample of important research on this topic, highlighting the most important findings.

TABLE 1.1
Changes in On-Campus Experience, 1994–2004

Metric	1994	1999	2004
≥4 days on campus	84%	71%	77%
Avg. contact hours per week	17.6	17.1	16
No collaboration with other students	30%	50%	60%
Engaged in part-time employment	48%	60%	60%
>10 hours paid work per week	40%	52%	49%

A large meta-analysis of class attendance and its relationship to student success was carried out by Credé et al. (2010). Analysis of 69 articles that reported a correlation between final grade, GPA, and attendance was conducted, and showed that class attendance was consistently one of the higher predictors of student success. Analysis of results from the 1970s through to 2010 also suggested that this relationship has not significantly changed over time, with a similar moderate correlation present across four decades. Four papers from the 1930s were also included in the analysis, once again showing a very similar positive correlation between attendance and GPA. The average correlation coefficient was 0.44 for all classes, 0.49 for science-related classes, and 0.41 for overall GPA. Although attendance rates were found to be correlated to both hours of study and previous academic achievement (SAT and high-school GPA), these correlations were below the level of that between attendance and grades. This indicates that attendance has an independent effect on grade and GPA. This study also investigated factors that influence attendance rates in class, and found that conscientiousness (self-reported), hours of study (self-reported), and high-school GPA had the highest positive correlation to attendance rates, while hours of work had the largest negative correlation. There was also a significant difference between genders, with female students having higher overall attendance rates than males, with correspondingly better grades.

Durden and Ellis performed a study in 1995 to investigate the link between attendance and results, focusing on economics classes at a single institution over three semesters (Durden & Ellis, 1995). Attendance data was self-reported, and showed that, while missing a small number of classes had very little impact on the final grade, there was a significant drop in grades for those students missing a large number of classes.

In order to more accurately quantify the effect of missing particular classes on student performance, Marburger looked not only at the effect of attendance on overall student performance, but also tracked the performance of students on particular topics based on their attendance in the classes in which that topic was discussed (Marburger, 2001). Students who did not attend had access to all materials online, including those materials relevant to each exam question. Average absenteeism over the semester was 18.5%, with a noticeable increase in the later weeks. The rate of absenteeism was almost 50% higher for Friday classes compared to the other days, which were roughly equal. The effect of missing the class that was relevant to each exam question was measured, while controlling for all other factors, and showed that missing a specific class related to an exam question had a statistically significant negative effect on the probability of a student answering that question correctly, increasing the chance of a wrong answer by approximately 14%.

In 2005, Gump conducted a study on a moderate cohort of students at a US university (Gump, 2005). The overall results and number of absences for each student were tracked over each semester, and the data analysed to determine any correlation. They showed a significant correlation coefficient of −0.601 between average grade and number of missed classes. Furthermore, in each class except one, students receiving an "A" had the lowest mean number of missed classes (0.61 overall), while students receiving the lowest passing grade of "D" had the highest (5.5 overall). Failing grades were excluded from this analysis. Female students had a higher overall attendance rate (0.94 absences) compared to males (1.41 absences), with a correspondingly higher overall grade.

A 2005 study by Kirby of attendance of economics students at an Irish university investigated the effect of attendance on a sample of 368 first-year students across two separate courses. The average attendance rate for these students was low, averaging 46% across both courses, with only a small number of students (4% and 2%) attending every class. Attendance rates were slightly higher in tutorial classes when compared to lectures. The results of study showed that travel time and hours worked were the most significant predictors of student attendance, with students travelling more than 30 minutes having higher rates of attendance, and hours worked being negatively correlated with attendance. This was thought to be due to students who lived further away staying on campus for entire days, while those living closer may opt to return home between classes. Female students also had a considerably higher rate of attendance than males, particularly at tutorial classes. It was also found that tutorial attendance had the higher overall effect on final grade, with lecture attendance having a smaller impact. Previous student performance, as measured by university entrance score, was also significant.

To overcome the limitations of self-reported attendance data, Kassarnig et al. (2017) used location and Bluetooth data from smartphones to create an objective metric of class attendance. Data was collected over a two-year period and included GPS location, proximity of other students via Bluetooth scan, and mobile phone communication logs. Social connections between students within classes were also estimated based on text message exchanges. The results of this analysis once again showed a moderate positive correlation (0.255) between attendance and student performance. Attendance rates showed a monotonic increase at each level of achievement. Interestingly, students with low attendance rates showed an almost uniform distribution of final grades, with results tending to skew highly positive as attendance rate increased. The top two quintiles of students attending almost all classes had both the highest mean and median grades, and a very low rate of failure. There was little discrimination between these quintiles, indicating that missing a small number of classes did not have an appreciable effect on final grade distributions. Attendance rates tended to fall over the course of a semester, with low-performing students showing the greatest decline. The second aspect of this study looked at how social groups affected class attendance. By monitoring text message communication between students, peer groups were identified, and the relationship between attendance of each student and the attendance of their peers was analysed. This analysis showed a moderate to high correlation between these variables, indicating that students were more likely to attend a class if their social group also attended. This observation has significant importance for online learning, as there is limited social interaction in such classes.

Social interaction with other students is also recognised as being a key contributor to student success and development. Previous students have found strong links between social interaction during university and GPA, retention, graduation rate, and career success, particularly among students with a lower level of previous academic achievement (Pascarella & Terenzini, 2005). Other research has shown that not only do on-campus social connections, including those maintained through online channels such as Facebook, have a positive impact on these factors, but also that a higher rate of similar social activity with friends from off-campus has a correspondingly negative impact, particularly within minority groups (Fischer, 2007). This work highlights the critical nature that peer friendships have on university success. Further research has shown that the nature of this relationship is complex, that particular aspects of friendship and social interaction are more important in encouraging campus attendance than others, and that both students and educational institutions must make an effort to facilitate such interactions (Braxton et al., 2013). A 2019 study evaluated the effect of friendships with a group of 2,739 US university students from multiple institutions and varied racial backgrounds, tracking GPA, graduation status, and survey data on their four closest relationships at the end of their second year of study (sophomore year) (Bronkema & Bowman, 2019). Four relationships at the end of second year was chosen due to previous research showing that three to five is a typical range for the number of people who provide the most emotional support (Dunbar, 2010), and that such relationships at university are typically not cemented until after two years of study (Newcombe & Wilson, 1966). The results of this study, controlled for other variables, showed that those students who reported having no close on-campus friends had significantly worse graduation and GPA outcomes, with almost *double* the rate of non-graduation. This is despite this group also reporting higher rates of study hours per week and lower average hours spent partying, both of which are independently associated with *improved* outcomes. The high-school GPA of the groups was not significantly different. Furthermore, having more close friendships on campus was found to further increase the chance of successful graduation. These results demonstrate the profound importance of developing close relationships with peers in a university setting, reinforcing previous work in this area (Goguen et al., 2010). However, it has also been shown that students are less interested in extra-curricular and other social activities on campus than they previously were, with a ten-year study showing that the number of students who were interested in extra-curricular events dropped from 44% in 1994 to 37% in 2004 (Krause et al., 2005).

WHY STUDENTS SKIP CLASS

Research has shown that the time students spend on campus is affected by many factors, including part-time work, family responsibilities, and other social and personal duties (Tinto, 2009); however, to date there have been relatively few studies that attempt to quantify this data in a meaningful way, making it difficult to track the changes in the reason for non-attendance over time. Some of the research providing the most insight into this topic is summarised here.

In a 1992 study of class attendance, Van Blerkom investigated attendance in psychology courses, showing that attendance typically decreased over the course of a semester, dropping from 93% in the first two weeks to 82% in the final weeks (Van Blerkom, 1992). The reasons for non-attendance were most frequently stated to be other required coursework, lack of engagement, illness, and social activities. A study undertaken at a Scottish university showed similar results, concluding that employment commitments, assessment for other courses, and illness were the top three reasons for missing class (Paisey & Paisey, 2004). However, the data also suggested that the time of the class was a large factor, with morning classes and those scheduled for later in the week showing significantly higher rates of non-attendance. Timetabling issues, personal problems, and lack of motivation were also cited as reasons for non-attendance. As suggested by previous research, the performance of students was positively correlated with attendance rates.

A study carried out at a London university with a sample of 155 students showed somewhat similar results (Kottasz, 2005). Illness (72%) and other coursework (61%) were listed as the top reasons for missing lectures; however, the reason "work commitments" (14%) was comparatively low on the list. Instead, scheduling issues (50%), other methods of obtaining content (38%), and poor-quality content and/or delivery (23%) were the three next highest reasons for not attending. Analysis of the reasons to attend class show that both extrinsic motivations, in this case scoring higher grades, and intrinsic motivation, expressed as a strong interest in the subject material, were the most important reasons for students to attend. Importantly, it was found that low motivation was a key factor in predicting non-attendance.

A study by Kelly (2012) investigated attendance at science-based courses over a two-year period (2007–2008) in Dublin, Ireland. This survey showed that class attendance was generally quite poor, with overall attendance rates of only 56% across the two-year period. Some interesting results of this study were that attendance rates varied significantly across the week, with Friday reporting a significantly higher rate of non-attendance. Despite students listing "too early" as an important factor in why they skipped class, time of day did not have a significant impact on attendance, with early morning classes showing similar rates to those in the afternoon. The study also showed that timetabling also impacted student attendance rates, with single classes on a day having lower attendance than those that were scheduled consecutively with others. Classes in the later years of study also showed a higher attendance rate, and it is proposed that this is due to increased motivation for students in more specialised classes later in the degree program. Travel distance was found to be only marginally correlated, with similar rates of attendance for those living on- and off-campus, and similar rates for those in the first and last quartiles of travel time. On-campus students did have slightly higher rates of attendance for early morning classes, which aligns with the survey results showing that social activities the night before had a lower impact on class attendance for these students (22%) when compared to those living off-campus (35%).

Allen and Farber (2018) investigated the reasons for non-attendance, in particular how travel and transportation options limit students' opportunities to attend campus. As the relative cost of accommodation either on-campus or close to campus has increased, students are increasingly living in locations that are either further from campus or have less access to fast and affordable transport to campus. Low access to transportation has been previously linked to reduced activity participation and an increase in social exclusion (Lucas, 2012; McCray & Brais, 2007), and this work aimed to determine if the same relationship was true for higher education. Data for the study was comprised of GIS data from multiple sources in a large metropolitan area (Greater Toronto), survey data collected from over 15,000 students, and the detailed travel diaries of 2,000 of these respondents. Results of a regression analysis showed that there was a strong link between travel duration and attendance rates, with those students requiring longer trips and transfers on public transport less likely to attend. This study also confirmed the previously identified relationship between hours of paid employment and attendance rate, with those students undertaking more work having significantly lower rates of attendance.

EFFECT OF ONLINE CONTENT ON ATTENDANCE RATES AND STUDENT PERFORMANCE

The rise of digital learning, and the ease of availability of online content, represents perhaps the greatest transformation in university education in many centuries. Previously, students had two main mechanisms of collecting information: in-person class attendance, and printed materials (usually a textbook). With online materials now ubiquitous, and most lectures recorded for later viewing, the reliance on these "traditional" forms of content delivery is reduced. By the early 21st century, students were consistently relying on the web for study materials and other purposes, with studies showing students spent an average of 4–5 hours per week studying online, with only a small fraction (3%) not using the web at all for learning purposes (Krause, 2005). Students were no longer required to spend time in the classroom, a physical library, or indeed on campus, to access most course materials. The impact of this technological shift on class attendance is, therefore, an important consideration.

Many studies investigating how the availability of online content has affected class attendance rates and student performance have been carried out over the last two decades. A 2005 study of 178 students investigated the relationship between use of online notes and exam performance, and also the effect on attendance rates (Grabe et al., Douglas, 2005). The results of this study found that, while the availability of online notes was a factor in over half of student absences, students who accessed online notes also performed slightly better during examinations. No analysis of the effect of attendance on student performance was undertaken, so it is not possible to know how students who used online notes and did not attend class performed relative to the cohort average.

The availability of online recordings of lectures has also been shown to have an impact on student attendance rates. A 2010 study showed that, while attendance rates dropped when online recordings of lectures were available, the effect was not as pronounced as when online notes were available (Traphagan et al., 2010). Furthermore, students who accessed these online recordings did not suffer the same reduction in overall grade that is predicted by non-attendance, and increased use of recordings was correlated with improved performance among both those who did and did not attend class in person. This indicates that, if used properly, lecture recordings are a valuable tool for improving student performance. The general attitude of students towards recorded lectures is also positive, with the majority viewing them as a useful tool in their study for both accessing missed classes and reviewing material prior to examinations.

A 2012 study reported interesting findings regarding the availability of online materials and assessment submission (Kinlaw, Dunlap, & D'Angelo, 2012). This study found that attendance for classes that offered online materials was *higher* than other classes at the same institution, with the average number of missed classes approximately 10% lower. This seems counterintuitive, particularly considering that the availability of online materials was stated as being a significant factor in the decision to miss class by more than half of surveyed students. Although no longitudinal data was collected, faculty members indicated that they experienced either no difference in attendance rates (94%) or a decrease in attendance (6%) after switching to online materials. No faculty members reported an increase in attendance. One possible explanation for this seeming incongruency is that those faculty members offering online materials, including online submission of assessment and course notes, were providing a higher-value overall educational experience for students, which research suggests is among the largest predictors of attendance rates (Dolnicar et al., 2009).

CLASS AND CAMPUS ATTENDANCE – CONCLUSIONS

This survey of the literature reveals a few important trends regarding student attendance. Despite the claims of many faculty, attendance rates at most universities remained relatively static from the 1970s until the mid-1990s, with any significant drops either temporary in nature or limited to local areas. The research shows a strong correlation between attendance and student success; however, the link is not always direct, with many confounding factors. Missing out on the on-campus experience also limits social activities and the possibility of group participation, which is also shown to adversely affect results. Reasons for non-attendance have also remained relatively consistent over this period, with students reporting illness, employment, travel time, and poor scheduling as the main reasons for not coming to class. Objective attendance data shows that this is not entirely accurate, revealing that absences increase significantly later in the week, for early morning classes, and later in the semester. As it is unlikely that students suffer more illness or work more at these times, other factors such as motivation and engagement are no doubt large contributors to these cases of absenteeism.

The advent of the digital age did see a significant drop in attendance, particularly as universities began to make some or all resources available online. This appears to have occurred between 1995 and 2000, which aligns well with the widespread availability of fast internet access and the shift to digital resources at most institutions. Despite the continued move to online resources, attendance rates have not fallen significantly since 2000 and, in fact, appear to have rebounded in some instances. Furthermore, the availability of online resources seems to have reduced the impact of absenteeism on student grades, although the correlation is still significant.

Ultimately, student attendance rates will be primarily affected by the value they place on attending classes. This "value" proposition is, in turn, determined by many factors – the perceived impact on final grade, their interest in the material, the knowledge and skill of the instructor, and the impact that attendance will have on activities that are competing for student time such as employment, family responsibilities, and social events. While class attendance is a powerful predictor of student results, it is equally clear that students do not always see value in this when compared to the negative aspects of attendance such as increased travel time, loss of income, and boredom. The availability of online lecture materials certainly appears to have negatively affected attendance rates and overall campus participation, with a marked drop from the mid-1990s through to 2000 being evident in most studies. Rates have since appeared to stabilise somewhat, and there is evidence that the higher availability of online materials has reduced the negative impact somewhat on grades from missing classes, particularly when both online notes and lecture recordings are made available.

To increase student participation in classes, both in-person and online, this correlation between attendance and results is insufficient as a motivator, and other barriers to attendance must also be removed where possible. Although some of these barriers such as employment and family responsibilities are outside of the control of educators, others are certainly able to be addressed. Class schedules can be improved to reduce travel time and eliminate single-class days. Teaching quality, regardless of the mode of delivery, can and should be improved to make the material more engaging and interesting to students.

THE CHANGING PREFERENCES AND EXPECTATIONS OF STUDENTS AND FACULTY

From the previous sections, it is clear that university education has significantly changed over the past 50 years, with both student demographics and the mechanisms of learning being vastly different to what was previously considered to be normal. It is, therefore, inevitable that current students and faculty alike will have different expectations of the university experience to those of the past. Despite the importance of identifying and adapting to changing expectations, this element of higher education has remained relatively under-researched to date (Wieser et al., 2017). In this section, we will look at some of the ways in which student and faculty expectations have changed, particularly in light of the recent trend towards online learning.

Students now expect their university education to fit within the rest of their lives, rather than vice versa (McInnis, 2001). Although the term "disengagement" is probably not entirely accurate as it implies that the students are lacking in some way, meeting the needs of these students is certainly a more difficult proposition than it has previously been. Greater competition between institutions in attracting and retaining students has also created a stronger bargaining position for students, as they can, and often do, simply "go somewhere else" if their needs are not being met (Darlaston-Jones et al., 2003). Students are also increasingly seeing a university education as a means to an end, rather than as a life experience. This is demonstrated by a 30-year longitudinal research of study attitudes carried out in 1996, which showed that students' interest in developing a "meaningful philosophy of life" dropped from 84% in 1966 to 42% in 1996, while their desire to "be very well-off financially" increased from 44% to 73% over the same period (Astin, 1998). Other changes in this time include a precipitous drop in political awareness, with only 30% of students regarding this issue as "important" in 1996, compared to almost 60% in 1966. In contrast, other studies have reported that students view interaction with both lecturers and other students as important, with a strong preference for smaller and more interactive classes over large lectures (Kandiko & Mawer, 2013).

Meeting the needs of students can also be quite challenging for universities. A survey conducted at an Australian university prior to commencing first-year studies, and then again at the completion of the year, showed that on almost every metric the university and teaching staff fell short of student expectations (Darlaston-Jones et al., 2003). Students rated facilities, equipment, reliability and dependability of teachers, sympathy, trust, and university support of teaching staff all below expectation, with trust scoring particularly far below expectation. Indeed, the only areas in which the university scored higher than expectation were in the dress standards of academic staff, and their understanding of student needs.

Sander et al. conducted a study in 2000 that investigated students' preferences and expectations of higher education institutions, by surveying almost 400 first-year undergraduate students (Sander et al., 2000). Results of this survey showed that while students expected to encounter mostly formal lectures during their studies, their strong preference was for a more interactive lecture experience, along with student-centred learning and tutorials. The least-wanted aspects of education were formal lectures, presentations, and role-play scenarios, although somewhat surprisingly private study also ranked highly in this question. Regarding assessment, there was a slight preference for coursework-based assessment over examinations; however, this varied by discipline, with medical students in particular favouring examinations. Breaking down the coursework assessment further, the most preferred forms were essays and research projects, with journals and presentations the least preferred. When asked "What makes a good teacher?", the most frequent response was "teaching skills" by a large margin, followed by knowledge and approachability. Interestingly, "organisation" was the lowest-ranked characteristic. Overall, the study demonstrated that students have a preference for styles of teaching and assessment with which they are already familiar over those that they may not have previously experienced.

Timetabling has also been shown to be a common area of concern for students, where the reality does not match their expectations. A 2017 study that surveyed students after one year of university and provided qualitative analysis of their responses found that, although students generally perceived the university as providing adequate levels of support, timetabling was a common area of concern (Money et al., 2017). Students generally would prefer classes to be timetabled on the same days to allow flexibility in outside employment rather than spread across the entire week.

A 2018 study in a Latvian university investigated student preferences and expectations of the physical environment (Licite & Janmere, 2018). It found that students had high expectations around technical facilities such as wi-fi access and interactive teaching facilities. Students also placed a high value on the comfort of the physical environment, particularly the chairs. However, when asked to choose between quality of teaching and quality of technology, teaching quality was overwhelmingly preferred. The overall results of the study conclusively showed that students want "good teaching, in a technologically enabled and comfortable environment". Other factors such as room layout, room size, and other factors related to the physical environment were of lesser importance to the students. This aligns closely with other studies that have shown that modern students expect good access to technology, and a comfortable environment (Kahl, 2014).

Student preferences for online learning have been difficult to predict and measure, with studies showing conflicting results. A 2018 study involving 107 students from

Taiwan showed a slight preference for in-person learning over online; however, the difference was not large and was within the margin of error of the survey (Bali & Liu, 2018). A 2018 study involving 165 business students in the US investigated student preferences for online versus "traditional" or blended courses and teaching (Weldy, 2018). It found that the majority of students (76%) preferred traditional in-person teaching, with only 12% preferring fully online courses. An overwhelming majority (87%) believed that they learned more effectively in traditional courses, while most (69%) thought that they received higher grades. While 88% of respondents believed that online courses required more self-learning, 59% thought that they did more study overall for traditional courses. It should be noted that students also showed a preference for podcasts and videos compared to discussion forums and traditional homework problems.

A qualitative study carried out immediately prior to the COVID-19 pandemic, and the ensuing mass migration to online education, looked at the expectations of online students from an Australian university (Henry, 2020). In-depth interviews were conducted online with 47 students from a broad range of disciplines and backgrounds, and the responses were analysed thematically. Motivation was found to be a common theme, with most respondents showing a high intrinsic motivation for study, and had high confidence of success, although there was concern regarding managing conflicting time commitments. Time management was seen as an essential skill that online students would need, given the lack of on-campus interaction. Time flexibility was seen as a key advantage to online learning, as many students had employment, family, or other responsibilities that would limit their ability to attend on-campus classes at normal times. Additionally, there was a common expectation or hope that employers, family, government, and others would provide support during their studies. Most respondents expected that the majority of their learning would be independent; however, they had an expectation that help would be provided promptly if it was required, through email and forums. Some level of organised peer discussion was also a common expectation. In terms of the learning experience, the expectation was that online students would receive a similar experience to those students studying on-campus, with the only difference being the mode of delivery. The content, assessment, and other activities were expected to be identical. Accessibility of technology was another common expectation, with students expecting high levels of reliability, and that all materials will be available promptly when required.

Since 2020, the number of classes being taught in online and blended mode has increased substantially due to the COVID-19 pandemic. While this has significantly changed the learning landscape in higher education, it has also provided an excellent opportunity to study the impact of enforced online learning on multiple cohorts of students. This research has shown a lot of variability in results, from a strong preference for traditional in-person teaching to a reliance on online and distance tuition. A 2020 survey of 130 IT students in Sri Lanka concluded that students have a strong preference for blended learning (54.7%) compared to in-person (28.1%) or online only (17.2%) (Akuratiya & Meddage, 2020). This survey also showed that students more frequently used mobile devices such as phones and tablets, and more often used mobile data sources instead of wi-fi to access online content and lectures. A similar study with responses from 488 medical students showed a very strong preference for in-person teaching (67%) over virtual (Al-Azzam et al., 2020). However, analysis of the results also showed that students who attended more online classes and engaged more with the content had an increased preference for these classes. Better access to online resources also showed a strong correlation with preference for online learning, and students who preferred online learning were also more likely to see an increase in their GPA since switching to online. A similar study examined the preferences of 247 medical students from Kansas, surveying students who had undertaken both in-person and virtual laboratory sessions (Brockman et al.,2020). Although most students preferred an in-person session (65%), those students undertaking the virtual lab reported that it was more convenient (66% vs 49%), and conveyed more new information. Conversely, those students attending the in-person session reported slightly higher rates of enjoyment (64% vs 59%). Overall, most students expressed a preference for a blended approach to laboratory sessions; however, students in each group also expressed a slightly higher preference for the mode of learning in which they participated, indicating the possibility of familiarity bias.

An important characteristic of the COVID-19-induced transition to online learning was the enforced nature of the switch. Unlike those students that made a choice to undertake online learning, both faculty and students alike were thrust into a foreign learning environment for which they were largely unprepared. For example, students who had previously relied heavily on mobile devices such as phones and tablets for online resources soon discovered that these were not sufficient for fully online study and assessment (Tick, 2019). A study of the enforced online experience of 2020 was carried out with 512 students from universities in South Africa, Hungary, and Wales who were previously enrolled in fully on-campus programs of study. This study found most students missed the in-person interaction, and generally had appropriate equipment, internet access, and environment to access the online materials. The major difference in responses between the three groups of students were regarding student engagement. While the students from the Hungarian university generally preferred the online environment and found interaction with other students and instructors to be easy, the opposite was true for students from both Wales and South Africa. It was also found that younger students had more difficulties in adapting to online learning than those who were older.

FACULTY EXPECTATIONS

While student expectations and preferences surrounding university in general, and online learning in particular, are complex, the attitudes of faculty are also important. Given that student interactions are far more limited in scope for online courses, the design and delivery of content is of even greater importance, with the potential for a lack of engagement and poor outcomes if these elements are not properly considered (Boling et al., 2012).

A 2008 snapshot of online education in the US gives a broad perspective of this, by surveying both teaching faculty and senior academic officers on the topic of online learning (Allen & Seaman, 2008). When asked what factors motivated online teaching, these two groups both ranked "meeting student needs for flexible access" and "the best way to reach particular students" highly. However, in the five remaining categories there was a large disconnect between teaching faculty and senior academics, particularly in the responses "to earn more income", "professional and personal growth", and "for pedagogical advantage". Overall, senior academic staff were far more positive about all motivations for online learning, while teaching faculty were motivated only by student needs.

A 2016 study compared the expectations of students ($n = 237$) to what faculty ($n = 71$) assumed students would expect prior to the semester, with the results showing that students generally had higher expectations than predicted by faculty in most areas of learning (Borghi et al., 2016). In particular, students had significantly higher expectations that they would be involved in community and voluntary work, that all course content and information would be available online, and that courses had a high level of requirements. The only areas in which faculty had higher expectations than students were those involving bureaucratic processes, services, and financial support. Overall, 19 of the 32 variables showed a significant difference between the level of student expectation and what faculty predicted, suggesting a knowledge gap between students and professors in relation to student expectations of higher education.

More recent studies in faculty perceptions of online learning show that a gradual shift in the way educators view online teaching has changed. Rather than a simple transformation of existing material into an online classroom, faculty now better understand that good online teaching requires a fundamentally different approach. A 2021 study carried out in a US university showed that most faculty viewed online courses as having fundamentally different requirements in terms of course organisation and communication in particular (Mellieon et al., 2021). Faculty recognised the need for multiple modes of teaching and communication to reach and engage with the greatest proportion of students, reporting that only with good planning and a solid understanding of the technical tools available could this be achieved. These findings are mirrored by another study at a Portuguese university, showing a generally positive attitude towards online teaching, while recognising the extra time and professional development that is required for effective teaching (Martinho et al., 2021).

DISCUSSION

In line with changes in demographics and attendance rates, the expectations of students in higher education have also changed considerably over the last five decades. While, in previous generations, students expected to be on campus for most of their week, the modern student expects that the university experience will be flexible enough to cater to their other commitments such as employment and family. Modern students are also both more knowledgeable about what they want from their education, and more direct in how they expect it to be supplied. There is an expectation that a university will provide precisely the skills that students need to be successful in the workplace, rather than a well-rounded, classical education. Despite an expectation for materials to be available in many forms for easy access, students also expect an interactive learning experience, and personalised assistance when encountering problems. The expectations surrounding university facilities have also changed, with students demanding a more modern experience and comfortable learning environment.

The introduction of online learning as a significant mode of delivery has further complicated the situation. Studies have consistently shown that the preference for online and in-person learning is extremely varied, making it impossible for the "one size fits all" approach to higher education that was previously dominant. The broad demographics of today's student body compounds this issue further, creating an impossibly disparate range of expectations that are often completely the opposite of each other. In delivery of material, lectures, once the cornerstone of the university experience, are now considered to be among the least desirable and engaging forms of learning by students and faculty alike, with online videos, interactive student-led sessions, and practical classes much preferred. Overall, universities are perceived as being slow to respond to changing student expectations, and somewhat out of touch. This is backed up by research showing that university management typically have a view of what students want that differs significantly from reality, leading to methods of teaching and university programs that are unappealing to all. There is also a somewhat cynical view that university management make decisions based more on financial incentives than pedagogical merit.

The response of higher education to the COVID-19 pandemic has led to somewhat predictable reactions from students and faculty. Prior to the pandemic, online education presented a new opportunity for students who were otherwise unable to attend campus, or preferred to study remotely. Similarly, online classes were offered by only a small percentage of academic faculty, with the majority of instructors being limited mainly to on-campus modes of delivery. Thus, the choice to undertake online study was largely internally motivated. With enforced online learning due to lockdowns and other responses to COVID-19, this was no longer the case. Students who preferred and benefited from on-campus experiences were no longer given that option, and staff who had no experience in delivering courses online were now forced to do so, with little or no training or preparation. This has inevitably caused the vastly different opinions of online learning since 2020, with some institutions reporting very high levels of student satisfaction, and others very low. Within individual cohorts and classes this pattern is often repeated, with some students enjoying the transition to online delivery and others finding it difficult to engage and learn successfully. It is difficult to draw any firm conclusions from the research to date, as not enough time has passed since the "mass migration" to online learning to assess the full impact. Just as it took a full decade for the higher education sector to react to the online world in the mid-1990s, it will similarly take time for both staff and students to adapt to the "new normal" of full or partial online learning. This will no doubt be a promising area of research in the next decade.

CONCLUSIONS

In this chapter we have attempted to provide a broad view of the current state of higher education, and how it has changed over time. Demographic shifts in recent times have been enormous, with increased access to higher education for females, minorities, and those of lower socio-economic backgrounds. With higher education participation rates rising sharply, the conditions of entry have also been reduced, meaning that universities are catering for students with lower academic ability, and lower levels of support. The era of the "full-time student" also appears to be coming to an end, with a majority of students now working either part-time or even full-time while studying. More students also juggle family commitments with study. The result of this change in demographics is particularly noticeable when looking at campus attendance. With more competing demands, students are often forced to choose between attending class and work/family commitments, meaning that absence rates are increasing, although not as quickly as the rate perceived by faculty. While this inevitably leads to poorer student learning outcomes, the rise of online learning and resources has somewhat lessened the impact by allowing students to access lectures and materials that they would otherwise have missed.

A thorough analysis of the available literature has also shown that the expectations of students have changed dramatically in recent times. Students are far more knowledgeable about what they want, and as well as having greater expectations, these are also far more varied. Students simultaneously demand more online content and more in-person interaction, more group work as well as more individual work, more self-study, and more opportunities for participation. These expectations are frequently misaligned with the expectations of faculty and university management, who instead tend to view their students as a more homogenous group with a single "preferred learning environment" that faculty should strive to create. Another key finding of the literature is that, regardless of what learning style is used, it is of critical importance that it is done well. Learning and teaching in online environments are considerably different and require both different tools and different skills to teaching in-person classes. Expecting faculty to adapt to this new style of teaching without adequate opportunity for professional development has conclusively shown to lead to poor outcomes. This was particularly evident during the COVID-19 pandemic, with faculty transitioning to online learning essentially overnight with poor results, while those faculty and institutions with prior experience had greater success.

How higher education providers react to the significant fundamental changes of the last 50 years will be critical for the success of the sector. Adopting a single model of learning will inevitably create dissatisfaction among many students and faculty. Attempting to meet the expectations of all students simultaneously will be costly in terms of time, money, and resources, and impractical for all but the most well-resourced institutions. Finding a balance between these two extremes will be a challenge for university management and faculty alike.

REFERENCES

Akuratiya, D., & Meddage, D. (2020). Students' perception of online learning during COVID-19 pandemic: A survey study of IT students. *Tablet, 57*(48), 23.

Al-Azzam, N., Elsalem, L., & Gombedza, F. (2020). A cross-sectional study to determine factors affecting dental and medical students' preference for virtual learning during the COVID-19 outbreak. *Heliyon, 6*(12), e05704.

Allen, J., & Farber, S. (2018). How time-use and transportation barriers limit on-campus participation of university students. *Travel Behaviour and Society, 13*, 174–182.

Allen, I. E., & Seaman, J. (2008). *Staying the course: Online education in the United States, 2008.* ERIC.

Astin, A. W. (1998). The changing American college student: Thirty-year trends, 1966-1996. *The Review of Higher Education, 21*(2), 115–135.

Bali, S., & Liu, M. (2018). *Students' perceptions toward online learning and face-to-face learning courses.* Paper presented at the *Journal of Physics: Conference Series.*

Blanden, J., Goodman, A., Gregg, P., & Machin, S. (2002). *Changes in intergenerational mobility in Britain.* Centre for the Economics of Education, London School of Economics and Political Science.

Boling, E. C., Hough, M., Krinsky, H., Saleem, H., & Stevens, M. (2012). Cutting the distance in distance education: Perspectives on what promotes positive, online learning experiences. *The Internet and Higher Education, 15*(2), 118–126.

Borghi, S., Mainardes, E., & Silva, É. (2016). Expectations of higher education students: A comparison between the perception of student and teachers. *Tertiary Education and Management, 22*(2), 171–188.

Braxton, J. M., Doyle, W. R., Hartley III, H. V., Hirschy, A. S., Jones, W. A., & McLendon, M. K. (2013). *Rethinking college student retention.* John Wiley & Sons.

Breen, J. (2002). *Higher education in Australia: Structure, policy and debate.* Monash University, Australia.

Brockman, R. M., Taylor, J. M., Segars, L. W., Selke, V., & Taylor, T. A. (2020). Student perceptions of online and in-person microbiology laboratory experiences in undergraduate medical education. *Medical Education Online, 25*(1), 1710324.

Bronkema, R. H., & Bowman, N. A. (2019). Close campus friendships and college student success. *Journal of College Student Retention: Research, Theory & Practice, 21*(3), 270–285.

Cardak, B. A., & Ryan, C. (2009). Participation in higher education in Australia: Equity and access. *Economic Record, 85*(271), 433–448.

Chowdry, H., Crawford, C., Dearden, L., Goodman, A., & Vignoles, A. (2013). Widening participation in higher education: Analysis using linked administrative data. *Journal of the Royal Statistical Society: Series A (Statistics in Society), 176*(2), 431–457.

Crawford, C. (2012). *Socio-economic gaps in HE participation: How have they changed over time?* Institute for Fiscal Studies, UK.

Crawford, C. (2014). *Socio-economic differences in university outcomes in the UK: Drop-out, degree completion and degree class.* Institute for Fiscal Studies, UK.

Credé, M., Roch, S. G., & Kieszczynka, U. M. (2010). Class attendance in college: A meta-analytic review of the relationship of class attendance with grades and student characteristics. *Review of Educational Research, 80*(2), 272–295.

Darlaston-Jones, D., Pike, L., Cohen, L., Young, A., Haunold, S., & Drew, N. (2003). Are they being served? Student expectations of higher education. *Issues in Educational Research, 13*(1), 31–52.

Devine, J. (2013). *Attitudes and perceptions of first year engineering students.* Paper presented at the AAEE 2013, Gold Coast, Australia.

Devlin, M. (2013). Bridging socio-cultural incongruity: Conceptualising the success of students from low socio-economic status backgrounds in Australian higher education. *Studies in Higher Education, 38*(6), 939–949.

Dolnicar, S., Kaiser, S., Matus, K., & Vialle, W. (2009). Can Australian universities take measures to increase the lecture attendance of marketing students? *Journal of Marketing Education, 31*(3), 203–211.

Douglass, C. (2012). *Participation Rates in Higher Education: Academic Years 2006/2007 – 2010/2011.* Department for Business Innovation and Skills, UK.

Dunbar, R. (2010). *How many friends does one person need? Dunbar's number and other evolutionary quirks.* Faber & Faber.

Durden, G. C., & Ellis, L. V. (1995). The effects of attendance on student learning in principles of economics. *The American Economic Review, 85*(2), 343–346.

Fischer, E. M. J. (2007). Settling into campus life: Differences by race/ethnicity in college involvement and outcomes. *The Journal of Higher Education*, *78*(2), 125–161.

Goguen, L. M. S., Hiester, M. A., & Nordstrom, A. H. (2010). Associations among peer relationships, academic achievement, and persistence in college. *Journal of College Student Retention: Research, Theory & Practice*, *12*(3), 319–337.

Goldin, C., & Katz, L. F. (1999). The shaping of higher education: The formative years in the United States, 1890 to 1940. *Journal of Economic Perspectives*, *13*(1), 37–62.

Gopalan, C. (2016). The impact of rapid change in educational technology on teaching in higher education. *HAPS Educator*, *20*(4), 85–90.

Grabe, M., Christopherson, K., & Douglas, J. (2005). Providing introductory psychology students access to online lecture notes: The relationship of note use to performance and class attendance. *Journal of Educational Technology Systems*, *33*(3), 295–308.

Gump, S. E. (2005). The cost of cutting class: Attendance as a predictor of success. *College Teaching*, *53*(1), 21–26.

Henry, M. (2020). Online student expectations: A multifaceted, student-centred understanding of online education. *Student Success*, *11*(2), 91–98.

Howard, J. L., Bureau, J., Guay, F., Chong, J. X., & Ryan, R. M. (2021). Student motivation and associated outcomes: A meta-analysis from self-determination theory. *Perspectives on Psychological Science*, *16*(6), 1300–1323.

Hubble, S., & Bolton, P. (2021). *Mature higher education students in England*. House of Commons Library.

K12 Academics. (2004). History of Higher Education in Australia. https://www.k12academics.com/Higher%20Education%20Worldwide/Higher%20Education%20in%20Australia/history-higher-education-australia

Kahl, C. (2014). Students' dream of a "perfect" learning environment in private higher education in Malaysia: An exploratory study on "education in private university in Malaysia". *Procedia-Social and Behavioral Sciences*, *123*, 325–332.

Kandiko, C. B., & Mawer, M. (2013). *Student expectations and perceptions of higher education*. London: King's Learning Institute, 1–82.

Kaspura, A. (2019). *The engineering profession: A statistical overview*, 14th Edition. Engineers Australia, AU.

Kassarnig, V., Bjerre-Nielsen, A., Mones, E., Lehmann, S., & Lassen, D. D. (2017). Class attendance, peer similarity, and academic performance in a large field study. *PLoS One*, *12*(11), e0187078.

Kelly, G. E. (2012). Lecture attendance rates at university and related factors. *Journal of Further and Higher Education*, *36*(1), 17–40.

Kinlaw, C. R., Dunlap, L. L., & D'Angelo, J. A. (2012). Relations between faculty use of online academic resources and student class attendance. *Computers & Education*, *59*(2), 167–172.

Kottasz, R. (2005). Reasons for student non-attendance at lectures and tutorials: An analysis. *Investigations in University Teaching and Learning*, *2*(2), 5–16.

Krause, K.-L. (2005). *Understanding and promoting student engagement in university learning communities*. Paper presented as Keynote address: Engaged, Inert or Otherwise Occupied, 21–22.

Krause, K.-L., Hartley, R., James, R., & McInnis, C. (2005). *The first year experience in Australian universities: Findings from a decade of national studies*. Department of Education, Science and Training, Australian Government.

Licite, L., & Janmere, L. (2018). Student expectations towards physical environment in higher education. *Engineering for Rural Development*, 17, 198–1203.

Lucas, K. (2012). Transport and social exclusion: Where are we now? *Transport Policy*, *20*, 105–113.

Mantle, R. (2021). *Higher education student statistics: UK, 2019/2020*. Higher Education Statistics Agency, UK.

Marburger, D. R. (2001). Absenteeism and undergraduate exam performance. *The Journal of Economic Education, 32*(2), 99–109.

Marks, G. N., Fleming, N., Long, M., & McMillan, J. (2000). *Patterns of participation in Year 12 and higher education in Australia: Trends and issues. Research report. Longitudinal Surveys of Australian Youth.* ERIC.

Martinho, D., Sobreiro, P., & Vardasca, R. (2021). Teaching sentiment in emergency online learning – A conceptual model. *Education Sciences, 11*(2), 53.

McCray, T., & Brais, N. (2007). Exploring the role of transportation in fostering social exclusion: The use of GIS to support qualitative data. *Networks and Spatial Economics, 7*(4), 397–412.

McInnis, C. (2001). Signs of Disengagement? The Changing Undergraduate Experience in Australian Universities. Inaugural Professorial Lecture. Centre for the Study of Higher Education, University of Melbourne.

McInnis, C., James, R., & Hartley, R. (2000). *Trends in the first year experience in Australian universities.* Department of Education, Training and Youth Affairs, Australia.

Mellieon, J., Harold I, & Robinson, P. A. (2021). The new norm: Faculty perceptions of condensed online learning. *American Journal of Distance Education, 35*(3), 170–183.

Money, J., Nixon, S., Tracy, F., Hennessy, C., Ball, E., & Dinning, T. (2017). Undergraduate student expectations of university in the United Kingdom: What really matters to them? *Cogent Education, 4*(1), 1301855.

National Science Board, National Science Foundation. 2019. Higher Education in Science and Engineering. Science and Engineering Indicators 2020. NSB-2019-7. Alexandria, VA. https://ncses.nsf.gov/pubs/nsb20197/

Newcombe, T. M., & Wilson, E. K. (1966). *College peer groups: Problems and prospects for research.* Aldine Publishing Company.

Norton, A., Cherastidtham, I., & Mackey, W. (2018). *Mapping Australian higher education 2018.* Grattan Institute.

Office for National Statistics. (2016). *How has the student population changed?* https://www.ons.gov.uk/peoplepopulationandcommunity/birthsdeathsandmarriages/livebirths/articles/howhasthestudentpopulationchanged/2016-09-20

Paisey, C., & Paisey, N. J. (2004). Student attendance in an accounting module – Reasons for non-attendance and the effect on academic performance at a Scottish University. *Accounting Education, 13*(sup1), 39–53.

Pascarella, E. T., & Terenzini, P. T. (2005). *How college affects students: A third decade of research. Volume 2.* ERIC.

Prahlad, A. (2020). *Trends in higher education.* Hanover Research.

Productivity Commission. (2019). *The demand driven university system: A mixed report card.* Commission Research Paper, Canberra. https://www.ons.gov.uk/peoplepopulationandcommunity/birthsdeathsandmarriages/livebirths/articles/howhasthestudentpopulationchanged/2016-09-20

Romer, D. (1993). Do students go to class? Should they? *Journal of Economic Perspectives, 7*(3), 167–174.

Sander, P., Stevenson, K., King, M., & Coates, D. (2000). University students' expectations of teaching. *Studies in Higher Education, 25*(3), 309–323.

Snyder, T. D. (1993). *120 years of American education: A statistical portrait.* US Department of Education, Office of Educational Research and Improvement.

Tick, A. (2019). An extended TAM model, for evaluating eLearning acceptance, digital learning and smart tool usage. *Acta Polytechnica Hungarica, 16*(9), 213–233.

Tinto, V. (2009). *Taking retention seriously: Rethinking the first year of university.* Paper presented at the *Keynote address at ALTC FYE Curriculum Design Symposium,* Queensland University of Technology, Brisbane, Australia, February.

Traphagan, T., Kucsera, J. V., & Kishi, K. (2010). Impact of class lecture webcasting on attendance and learning. *Educational Technology Research and Development, 58*(1), 19–37.

Tuchman, B. W. (2011). *A distant mirror: The calamitous 14th century.* Random House.

Van Blerkom, M. L. (1992). Class attendance in undergraduate courses. *The Journal of Psychology, 126*(5), 487–494.

Weldy, T. G. (2018). Traditional, blended, or online: Business student preferences and experience with different course formats. *E-Journal of Business Education and Scholarship of Teaching, 12*(2), 55–62.

Wieser, D., Seeler, J.-M., Sixl-Daniell, K., & Zehrer, A. (2017). *Online students' expectations differ: The advantage of assessing students' expectations in online education.* Paper presented at the *Proceedings of the 3rd International Conference on Higher Education Advances.*

Theme 2

General Strategies, Approaches, and Practices for Online and Blended Learning Post-COVID-19

2 Engagement in the Blended Learning Model of a Post-COVID University

Anna Wrobel-Tobiszewska, Sarah Lyden,
Damien Holloway, Alan Henderson,
Peter Doe, and Benjamin Millar
University of Tasmania, Tasmania, Australia

CONTENTS

DOI: 10.1201/9781003263180-4

INTRODUCTION

The COVID-19 pandemic created an enforced pilot opportunity for a shift towards online teaching, which may be seen by universities as a more efficient and less costly teaching approach that is more convenient to students. During this time, we have also learnt a great deal regarding the benefits of face-to-face learning on the mental health of both students and staff, our interactions, and our sense of belonging to the university. Going forward, we assume that online teaching will play an increasingly important role in university education, as universities will relish the smaller footprint and infrastructure needs associated with the online classroom. We also assume that face-to-face learning will return in some capacity, though with a greater focus on quality interactions.

While some academics are already embracing these changes, the accelerated and inevitable shift to online and blended learning created by the COVID-19 pandemic has identified a group of engineering academics who are eager to adopt new practices but may not yet feel they have the skills to do so, and perhaps some others who are more reluctant but are willing to be convinced to change. This chapter is aimed at these groups, and is divided into three main sections. The first section discusses the elements of an ideal blended learning environment, including guidelines for different activity types such as asynchronous online videos and activities, and face-to-face sessions. The second section focuses on specific techniques for improving student–teacher engagement in online classes, adapted to suit an engineering context. These techniques are presented in three groups based on the amount of effort likely required for implementation, helping academics to make an immediate impact. Many of these techniques require the use of online tools, and we will present how these tools can be used to implement the introduced range of engagement activities. The third section provides strategies for encouraging colleagues to engage sooner rather than later in this inevitable change, informed from our own experiences. Our hope is that this chapter will empower engineering academics to lead the change within their institutions to move towards a more engaged engineering student cohort in the blended learning environment of a post-COVID world.

ELEMENTS OF AN IDEAL BLENDED LEARNING ENVIRONMENT

This section discusses the elements of an ideal blended learning environment, including guidelines for asynchronous online videos and activities, and face-to-face sessions. We discuss the importance of establishing online social presence by welcoming students in the online/blended learning environment, setting clear two-way expectations, and maintaining presence and engagement throughout the whole learning timeframe. Pre-recorded lectures, asynchronous learning activities, live-online sessions, and face-to-face activities are listed as the four most important components of modern blended learning environments. We summarise different approaches to setting up assessment items and providing feedback to students to accommodate the requirements of online learning and stimulate students' engagement.

WELCOME AND EXPECTATIONS

For a teacher, establishing online presence is the process of demonstrating their role in the online environment, and making social connections with others who share the environment. A well-accepted recommendation in the literature is that online teachers need to establish their presence from the first day. It needs to be welcoming and supportive with expectations clearly set out (Swinburne, 2021).

Welcome Students to the Unit

A good approach is to upload a short (2–6 minute) pre-recorded welcome video around one week before the semester starts. The video should show your face and welcome students to the unit, inform the students about what will happen during the first week of the semester, and the general expectations for the unit.

Introduce Teaching Staff

It is important for students to know who their lecturers and other teaching staff are. A lecturer plays a key role in the unit. The introduction may involve sharing information about your background, such as where you are from and some of your industry or research experiences. This helps to provide credibility in your role as a learning facilitator. This might be included in the welcome video or as a separate announcement or email to all students.

Establish Communication Channels

Students should know what the regular teacher-student communication procedures will be right from the start of semester. There should be clear information about how students can ask questions and when they can expect answers. Mutual expectations should be spelled out, outlining what students are expected to do during the semester, and what the lecturer is expected to provide.

Provide Assessment Task Details

The importance of having a logical, organised, and easy-to-navigate structure for online resources is crucial. The assessment tasks should include rubrics to explain both criteria and level of attainment.

Encourage Web Cameras during Web Conferencing

Castelli and Sarvary (2021) describe good reasons for not making webcams compulsory during web conferencing activities. Some students may not have access to private space or be embarrassed by showing their home environment, which may contribute to a negative student experience. Castelli and Sarvary (2021) suggest that lecturers should instead use positive encouragement by highlighting the benefits of participation in web conferencing, such as communication by nonverbal cues such as smiles, frowns, head nods, and looks of confusion and boredom. Nonverbal cues allow lecturers to modify their techniques during an activity to provide more effective teaching. Picking up a confused expression may prompt the lecturer to provide further explanation. The use of webcams also helps to build instructor-student and student-student relationships, which are known to reduce loneliness associated with remote learning.

ESTABLISHING AND MAINTAINING SOCIAL PRESENCE IN AN ONLINE ENVIRONMENT

The lecturer's social presence is very important in an online learning environment and needs to be actively managed to sustain students' engagement. Social presence in the classroom can be characterised for a teacher as behaviour that creates immediacy, such as verbal and nonverbal interactions, gestures, smiling, humour and vocal variety, personalising examples, addressing students by name, questioning, praising, initiating discussion, encouraging feedback, and avoiding tense body positions (Hackman & Walker, 1990). All these tend to happen more naturally in face-to-face interactions, but have to be planned and intentionally implemented in an online environment.

Social presence is positively related to student satisfaction with online courses (Leong, 2011). Aragon (2003) describes strategies for creating and maintaining social presence for instructors in an online environment, which are further explored below.

Actively Contribute in Online Discussion

Aragon (2003) describes a lecturer not actively participating in online discussions as equivalent to giving a lecture and leaving without interacting with the students. Discussion boards and other modes of asynchronous communication take the place of the many verbal interactions that would occur inside the classroom. Lecturers need to make sure they actively participate in these online forums. For example, it is good practice to refer to specific points that students have made in their posts. Regular and consistent contribution to these discussions will model and maintain effective online engagement.

Prompt Online Responses

Timely responses from instructors are invaluable to the establishment of social online presence. Aragon (2003) suggests that a good rule of thumb is to respond within 24 hours. Responses should be provided during reasonable working hours (8am–6pm Monday to Friday) and should not be sent at odd hours, to avoid creating an expectation from students that the lecturer will reply at any time of the day. Communicating an expected time for a lecturer to respond to student communication is a helpful way of managing student expectations.

Building Rapport by Adding a Human Component to Online Interactions

Many "human" interactions, including body language, occur naturally in a face-to-face environment. In an online environment, personal experiences on topics relating to the course, or humour, can be used. This has been found to greatly facilitate social online presence (Aragon, 2003). It shows the instructor's credibility in the subject matter and allows students to see their "human" side. Humour invites discussion, is a social leveller, and conveys goodwill, but needs to be used carefully online to avoid misinterpretation.

ONLINE STUDENT-LECTURER COMMUNICATION

Students often identify a preference for online communication with their peers due to fast response and ease of access, but face-to-face meetings are preferred when time permits (Merdian & Warrior, 2015). It is particularly important to be proactive about setting and maintaining online communication when teaching face-to-face and online simultaneously. Online communication between lecturers and students can use formal methods such as email, lecturer-initiated online discussions, or less formal methods such as online chat apps.

Class Announcements

Within the learning management system (LMS), class announcements can be used to share important information with a whole class. This form of communication however does not usually facilitate two-way or peer to peer communication, so additional tools may be required as outlined below.

Online Discussion Forums (ODF)

Online discussion forums are online fora with a tree-like structure of topics based on categories, each of which may include sub-fora. Topics in each sub-forum are called *threads*, where members can place posts. Uses of ODF can include providing feedback, forming small group discussions, or providing further discussion of a topic. The tree structure organises the discussion for the members to review specific topics later (Biriyai & Thomas, 2014).

Online Chat Apps

Students will generally communicate among themselves and form their own groups for study and social purposes, but online chats and/or surveys can be a useful way to enhance communication between students and student-lecturer communication. Many online group chat apps are available including WhatsApp, Discord, Slack, FB Messenger, Viber, Skype, Snapchat, WeChat, Telegram, Line, and Signal. When the instructor is present in a study or class group chat, moderation is necessary in order to maintain acceptable standards, and remove offensive or off-topic posts when required.

PRE-RECORDED VIDEO LECTURES

Best practice in online asynchronous delivery suggests pre-recoded video lectures should be in small, digestible segments of 10–15 minutes in length (Brookfield, 2015). A study from Massachusetts Institute of Technology showed that students, in the context of Massive Open Online Courses (MOOCs), typically would stop watching videos of 9 minutes' length around half-way through (Guo, Kim, & Rubin, 2014). Humphries and Clark (2021) demonstrated, via data of students downloading videos in first year units at an Australian university, that there was a significant preference for videos of 3–17 minutes' length over traditional length (i.e., 50-minute) lecture recordings.

Content Chunking

Smaller videos can be achieved by "content chunking" (Pappas, 2013). Content chunking breaks a topic down into sub-topics that can be viewed in a sequence to make better use of short-term working memory, and can be used as a strategy for developing small lecture videos. Figure 2.1 shows how a traditional lecture topic could be replaced with content chunking as a series of concept videos.

Interactive Videos

In addition to producing concept videos, tools like H5P (2021), which enables interactive HTML content to be created and embedded online, can be used to create interactive videos. Students might be asked questions throughout the video and their responses recorded to gauge their level of understanding, as would occur during a live lecture.

Concept videos

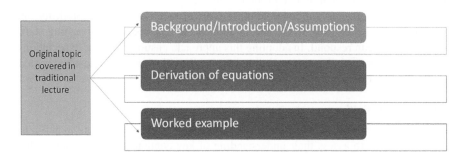

FIGURE 2.1   Replacing a traditional lecture with concept videos.

Existing Videos Resources

Resources on YouTube or other platforms may be useful additions to pre-recorded video resources for students, but it is important to be aware of the relevant copyright requirements, and links should be checked regularly as third-party material may be deleted or moved. The general rule is that the creator (author) is the first owner of copyright. The *Copyright Act 1968* gives creators the right of attribution and the right of integrity in their creation.

Other Video Types

In addition to turning lecture material into short, pre-recorded segments, one can also prepare a wide range of other videos for students, including Frequently Asked Questions (FAQ) videos, videos on assessment tasks, welcome videos, and feedback videos.

General Video Production Tips

When recording videos, ensure the environment is well set up to produce a professional video. It is best to be facing the light source. A good quality microphone in close proximity can substantially improve the audio quality. Eye contact is important to engage with the audience. To enable this, keep the camera level with your head (this may require adjusting the camera position) with a similar frame to a portrait photo and, while talking, look directly into the camera.

There are many online videos showcasing how to create effective and engaging online videos with simple equipment, which can be found by searching "How to make a video".

OTHER ASYNCHRONOUS LEARNING ACTIVITIES

Asynchronous activities are activities, usually online, that do not occur in a live class setting, including student emails, discussion boards, students completing work outside of tutorials, and students watching pre-recorded videos.

Asynchronous Consultation

Encourage student collaborative problem-solving by offering asynchronous consultation through a discussion board or other tools as mentioned above. Students can post questions about class work, and in addition to responses from teaching staff, other students can respond. In adopting this approach, ensure that clear expectations are set at the start of semester regarding when staff will respond to questions posted on these platforms.

Embedded Activities

Activities can be embedded in the LMS alongside pre-recorded lecture material to enable students to assess their understanding of key content and apply their skills. This is often used in a flipped classroom approach, and can identify areas where students need additional support within face-to-face interactive class time (Lo & Hew, 2019). HTML5 Package (H5P) may be available as a plugin to your LMS to create interactive HTML embedded activities. If not available, other tools within the LMS, including quizzes and surveys, could be used. For the best outcomes, these learning activities should be developed with the goal of providing students with formative feedback on their development of knowledge and skills, leading to achievement of the intended learning outcomes.

LIVE-ONLINE SESSIONS

There are many ways to enhance the delivery of live-online teaching activities to maximise student engagement and provide an intrinsic motivation to actively participate.

In a study of electrical engineering students, Petrović and Pale (2015) list the top five reasons for students choosing not to attend a lecture as:

- Lecture content can be mastered without attending.
- Poor performance by lecturer.
- Lecture content is not important to me.
- Course is not interesting to me.
- Lectures don't help me in comprehension of course content.

This study investigated face-to-face teaching, but the findings are equally relevant to online lectures, perhaps even to a heightened degree due to the additional challenges of keeping the audience engaged in a limited but largely uncontrolled environment – limited, as many aspects of face-to-face teaching are harder online; and uncontrolled since the audience is not captive and students are probably engaged in other activities at the same time.

Lecturer's Performance

In order to foster enthusiasm in a topic through teaching, it helps for the lecturer to be enthusiastic. A lack of enthusiasm will most likely "rub off" on students in a negative sense. Spend time researching and designing the lecture content in a way you are more comfortable with and find interesting. This will be more likely to attract a positive student response.

Online live lectures are particularly challenging. Often the audience is not visible, as students don't turn on their webcams. This makes it very hard to gauge student engagement and reaction during the class. Information communicated through facial expressions is no longer available. Instead, you will need to ask for comments and feedback, perhaps making use of synchronous chat, online polls, or other methods.

A final but critical aspect of online performance is the quality of audio – don't underestimate this. In any custom-designed professional recording studio you will see that professional announcers have the microphone only a few centimetres away from their mouths. Headsets provide reasonable sound quality at relatively low cost.

Student Participation

The more active the participation, the more likely students are to be engaged. Some things students might do in a typical traditional class, in roughly increasing order of mental activity required by the students, are listening (passively) to a lecture or video without taking notes; reading, taking notes while listening to a lecture; answering questions; solving problems alone; group discussions; and activities where the students are "put on the spot".

Activities where the students are "put on the spot" require significant mental alertness and task engagement, since the student must not only complete the assigned task, but is under pressure to perform at a higher level because they are being observed by their peers. Though it is less often used, it is easy to introduce in most online classes. For example, when solving a problem, pick a student at random to complete the next step. Then to put that student at ease and engage the rest of the class, invite the class to help the student, or use the "pass the chalk" technique described in the "Techniques for improving student-teacher engagement" section of this chapter. Ask leading questions, but resist the temptation to step in and do it.

Some examples of activities used to enhance student participation are presented in Table 2.1.

One good approach to keeping students focused on an online lecture is to keep them busy while maintaining two-way communication. When communication is one-way (as it often is), make sure you maintain the pace – don't leave gaps and pauses that can be filled with other things.

Relevance

The relevance of live-online sessions is important from the perspective of creating a bigger picture and creating an immediate connection between theory and application.

TABLE 2.1
Activities to Enhance Student Participation

Student Activity	Recommended Tools (Details Later in This Chapter)
Answering questions	Socratic questioning
Taking notes while listening to a lecture	Mind Dump
Activities where the students are "put on the spot"	Pass the chalk

---

A good introduction might make reference to case studies, but without too much technical detail. For example, a discussion of significant relevant structural failures can be used to highlight the importance of a topic in structural engineering. Many students find it difficult to absorb new theories, or indeed, even find the motivation to learn them, unless they can see examples and applications.

The following strategies described later might also help to highlight the relevance of a topic:

- Pre-empting topics to inspire students to attend the next instalment.
- Using students' personal experiences.
- Storytelling.
- Theme/statement/topic of the day.
- Ethical dilemmas.

Technology

Various tools and software described in the following sections can be used for live-online sessions depending on the requirements and expected learning outcomes.

ENHANCING FACE-TO-FACE SESSIONS (PRACTICALS, TUTORIALS, WORKSHOPS)

Face-to-face teaching will always be important, but modern universities are heading in the direction of increased blended delivery, with many of the COVID-related blended learning initiatives here to stay. Therefore, we must consider the role of face-to-face activities, how they can best be used to maximise educational value (rather that what is most expedient), and how online activities might support and enhance them. The latter includes both preparatory activities (such as pre-reading before a lab class), and follow-up activities (such as assessments or reflections), but might even be extended to synchronous activities (such as online polls or quizzes during a lecture).

Hybrid Synchronous Sessions

Hybrid delivery is delivering the same content at the same time to both online students and students attending in-class. Equipment normally available in an AV-equipped classroom includes a desktop PC, a data projector, and document camera; however, some rooms are equipped with a ceiling camera that follows the presenter.

A particular current challenge is how to simultaneously cater for the different needs of students who may be in the same classes but learning in person or online. For these classes:

- If the cohort is large, it may be possible to have separate sessions for face-to-face and online students.
- If separate sessions are not possible, the face-to-face session could be recorded or live-streamed, with an option for follow-up afterward with the online students. This will encourage engagement and identify any gaps in the information these students have absorbed.

Sharing Work on a Whiteboard

Work on a whiteboard can be shared with both online and face-to-face students using a portable document camera, rotated towards the whiteboard, to show the tutor addressing the students in-class and writing on the board. In-class students could also be linked to a video conference (such as Zoom) on their laptops or smartphones so that they can see shared content in high definition. This approach allows the tutor to move freely around the room to engage with in-class students. The portable document camera can easily be realigned to show the interaction between the tutor and in-class students, or to facilitate students' group work. If there is audio-visual equipment in the room, the desktop computer can be linked to the video conference and project the shared content.

ASSESSMENT TASKS

Blended and online delivery limits the effectiveness of many traditional forms of assessment. Some alternative options that may be considered are presented below.

Online Workshop-Based Assessment

During a live-online workshop session, students can be randomly allocated into breakout rooms; they can then submit the solution to the learning activity as an assessment item.

Online Discussion and Reflection-Based Assessment

If students are required to post within online discussion forums, their engagement can form part of an assessment task. Similarly, students may be asked to submit a reflection of key learning outcomes from pre-recorded lectures.

Dependency Links between Assessments and Learning Material

Most online learning environments have the capability to make the release of material and activities dependent on completion of prior tasks. For example, students can be required to solve web-based tests and problems before progressing to more advanced material.

Video Assessments

There are many ways in which videos can be included in the assessment scheme. Students can be asked to generate a group/individual video to address certain problems or answer a set of questions. An audio-visual explanation of working through the problem and/or presenting every step of working through equations can be used as an evidence-based video assessment. Another example of using videos as assessments can be a 1-minute message (1MM), where students are asked to record a 1-minute video synthesising material from a few pre-recorded videos. This technique can be used in take-home exams to help verify academic integrity.

Online Group Presentation

Students in the online environment can be asked to deliver an online group presentation that includes genuine live interactions between presenters. As this involves additional challenges not experienced in in-person group presentations, some guidelines on related skills should be provided to students prior to this task.

Peer-Review Assessments

In such assessments, students are not only asked to complete their own writing/problem-solving assessment component, but also to provide a constructive critique on a fellow student's assessment. According to the literature, peer assessment works best with more mature students (3rd–4th year) (Bates, 2019) but would be appropriate at lower levels if the weighting was not too high.

Online Peer Assessment

Peer assessment encourages collaboration and engagement. Feedback supplied from peer assessment tools can be automated, requiring attention from the lecturer at the setup only. Students' entries can be kept anonymous, unless otherwise required. It can also be used for students' self-assessment.

Weekly Contribution to an Assessment

The summation of small weekly tasks can form an assessment. For example, students could be asked to answer one question related to weekly topics and update a file shared with their lecturer. The lecturer would not necessarily be required to mark the answers every week; for example, they could audit engagement throughout the semester and mark the full task at the end of semester. This might potentially work better with smaller cohorts, unless an automatic approach to checking updates is used. Alternatively, there could be two rounds of feedback; e.g., in the middle of the semester, and at the end. This method would help maintain students' engagement throughout the whole course.

FEEDBACK

Written Feedback

Written feedback is the most common form of useful feedback. Feedback comments should concentrate on both positive and negative aspects of submitted work. Annotation of rubrics can be used to give students feedback on how an assignment was assessed.

Audio or Video Feedback

Audio feedback is easy to generate using commonly available software, especially within most LMS. It gives a student a more personalised approach, and it can save a lot of time – it would take minutes to write what you could say in 30 seconds of audio. Video feedback adds another layer to audio feedback and is particularly useful in blended or online units. It is suggested that at least one set of feedback should, ideally, be based on audio or video feedback.

Student–Teacher Feedback

It is also important to allow students to provide feedback to teaching staff. This could be on assessments and/or comment on problem solutions, and might be useful during semester so that staff can adjust delivery.

TECHNIQUES FOR IMPROVING STUDENT–TEACHER ENGAGEMENT

Building rapport with students has been identified as one of the crucial components of students' effective engagement in the online learning environment. This second section focuses on specific techniques for improving student–teacher engagement in online classes, adapted to suit an engineering context. These techniques are presented in three groups based on the amount of effort likely required for implementation and helping academics to make an immediate impact. It is a shortlist, culled from a list of over 100 techniques (Yee, 2019), and adapted to improve relevance to engineering units and to online teaching.

Many of the techniques require online tools. Table 2.2 lists some online tools that we have discovered, indicating which tools are suitable for each technique listed below.

Quick to Introduce

The ideas below require minimal preparation:

1. Discussion Stimulating Picture – Begin the lecture (or break the lecture) with an image meant to provoke discussion or emotion (another option: a cartoon). This could be an unusual use of some of the engineering components. To follow up, a question could be put next to it, and you could ask students to type a reply in the chat.
2. Picture Prompt – Show students an image with no explanation and ask them to identify/explain it using terms from the lecture, or to name the processes and concepts shown. This idea could be applicable to group activities, possibly held in breakout rooms in a video conference.
3. Socratic Questioning – The instructor replaces part of a lecture by peppering students with questions, always asking the next question in a way that guides the conversation towards a learning outcome (or major driving question) that was desired from the beginning. Variation: A group of students writes a series of questions relating to the topic given by lecturer as homework and leads the exercise in class (applicable in higher level years). This idea could be used when participating anonymously, by typing their answers or putting sticky notes on the whiteboard (see Table 2.2).
4. Recall, Summarise, Question, Connect, and Comment – This method of starting each session (or each week) has five steps to reinforce the previous session's material: recall it, summarise it, phrase a related unanswered question, connect it to the topic as a whole, and comment. To do this exercise ask for volunteers to share, or randomly pick one student. At the beginning of the semester, you should give students a heads-up that random callouts will be held during live sessions and keep track of the "selected" students. In adopting this approach, ensure that any required learning modifications for individual students are considered.

TABLE 2.2

How to Implement Recommended Activities with Common Tools

	Video Conferencing Software (such as Zoom)				Digital Interactive Tools (such as Mural/Jamboard)	Online Forms (such as Google forMS/Office 365)	Online Shared Documents (such as Google Doc, Office 365 Doc/Spreadsheets)	Quiz Software (such as Socrative, Slido)	Online Communication Channel (such as Slack)
	Annotate/WhiteBoard	Chat	Breakout Rooms	Polling					
Quick									
Discussion Stimulating Picture		×							
Picture Prompt		×	×						
Socratic Questioning	×		×	×	×				
Recall, Summarise, Question, Connect, and Comment		×				×	×		
Pre-empting topics to inspire students to come back									

Moderate	Using students' personal experiences	×		×		×		×	×	×
	Punctuated Lectures	×				×	×		×	×
	Student Polling						×	×		
	Storytelling									
	Reflection Break			×			×			
	Why Do You Think That?	×				×			×	×
	Mind Dump			×	×					
	Theme/statement/topic of the Day									
	Pass the Chalk									
Additional	Ethical Dilemmas Tournament		×					×		
	Online Recap				×		×			×
	Self-Assessment of Learning Styles						×			×

5. Pre-empting topics to inspire students to attend the next instalment – Rather than making each topic fit neatly within one day's class period, intentionally structure topics to end three quarters of the way through the time, leaving one-quarter of the time to start the next module or topic. This generates an automatic bridge between sessions and better meets the principles of effective learning resulting from interleaving topics. A good variation of this method to use in engineering units, is to intentionally inform students about this approach, and add a challenge to get students to think on how today's lecture applies to the next topic.

MODERATE EFFORT REQUIRED

These next activities will require some additional time to prepare:

1. Using students' personal experiences – Design your class activities (or assignments) to enable students to share their real-life experiences with the class. Instead of asking for reflections on a particular topic, ask students to voluntarily share personal experiences of a certain technical- or engineering-related subject. The experiences need not necessarily relate to engineering but may possibly open a different way of thinking by getting them to relate the experiences they know to the actual engineering experiences.
2. Punctuated Lectures – Ask students to perform five steps during a class: listen, stop, reflect, write, give feedback. Students become self-monitoring listeners. This provides a lot of value, especially for first year students. The writing part could be anonymous, or comprise sticking notes on a board (see Table 2.2).
3. Student Polling –Select some students to move virtually around the "room", polling each other on a topic relevant to the course, then report back the results to the class. Using an online web conferencing platform (such as Zoom), send students to breakout rooms, and nominate one person to poll others on understanding. The challenge for the lecturer is to come up with a question to check the students' level of understanding.
4. Storytelling – The lecturer illustrates a concept, idea, or principle with a real-life application, model, or case study. Personalise the delivery, possibly choosing examples that relate more closely to your own experiences or to case studies in which you have been involved.
5. Reflection Break – Ask a rhetorical or real question, then allow at least 20 seconds for students to think about the problem before you go on to discuss it. This technique encourages students to take part in the problem-solving process even when discussion isn't feasible. Having students write something down (while you write an answer) helps assure that they will, in fact, work on the problem. Keep it as an option for recurring questions throughout the semester, and potentially make the answers an assignment. Alternatively, add complexity and build on a topic by asking questions of a more complex nature.

6. Why Do You Think That? – Follow up all student responses (not just the incorrect ones) with a challenge to explain their thinking, which trains students over time to think in discipline-appropriate ways. Doing this online could be quite intimidating for some students; it would be imagined that this might put some students off responding, particularly if the session is being recorded. If they don't want to speak, they can type in a chat. Then don't give them the answer, but ask other students if they think the previous answer was correct.

7. Mind Dump – Students write for five minutes on a previous teaching activity, and the answers are collected. The entire chapter worth of mind dumps are returned at the end of the semester/module as a surprise to help students study for the test. Possibly use an online form (see Table 2.2) so that the lecturer has the actual names, but they are removed before making the summary available to all students.

8. Theme/statement/topic of the Day – Select an important term and highlight it throughout the class session, working it into as many concepts as possible. Challenge students to do the same in their interactive activities. This may have particular value for international students. For example, "Sustainable Engineering" in environmental engineering lectures, "Boundary layer", "Separation" in fluid mechanics, or "Stage of design" in design units.

9. Pass the Chalk – During tutorial-style online activities, students nominate the next person to answer the next question or solve the next presented problem.

ADDITIONAL EFFORT REQUIRED

These activities may require more time for the lecturer to establish:

1. Ethical Dilemmas – Incorporate case studies that include ethical dilemmas into the lecture. This has the advantage of providing links to real-world problems and adding to the technical aspects of the unit. Students could be asked to think about compromising between two different and potentially conflicting goals within a case study.

2. Tournament – Using breakout rooms, divide the class into smaller groups (at least two) and provide them with the opportunity to study together to review a topic before completing a quiz, tallying up the points. Then repeat with a second topic with study time, quiz, and tallying points. This could be useful as a game to review knowledge from previous units in a friendly competitive atmosphere.

3. Online Recap – Provide students with a form or outline where they can answer questions related to content recently covered within a particular section of a lecture. These could be collated automatically using tools in Table 2.2, allowing the lecturer to see where students may be having challenges understanding a topic. It could be useful for assessing student understanding during teaching of topics known to be challenging.

4. Self-Assessment of Learning Styles – Get students to complete a standard inventory of learning styles assessment (for instance – https://www.mint-hr.com/mumford.html) and share their overall learning style preference results (anonymous or not). This provides students with an understanding of their own learning style preferences and can help lecturers understand how to potentially update lecture materials and notes to better serve the learning style preferences of the cohort. Training in how to cater for different learning styles may also be beneficial for lecturers.

Table 2.2 presents a summary of the introduced engagement activities along with potential tools that could be utilised to implement these in an online environment.

STRATEGIES FOR ENCOURAGING COLLEAGUES TO ENGAGE

This third section provides strategies for encouraging colleagues to engage sooner rather than later in this inevitable change, informed from our own experiences. The section explores barriers, motivators, and positive actions.

BARRIERS

It is natural for lecturers to prefer the delivery method they have been using for face-to-face delivery. In many cases the delivery method mimics what they experienced as a student. One needs to be aware that personalities may align more towards either a fixed or growth mindset. The fixed mindset mitigates against change, but there is a lot of good recent literature on brain plasticity and on developing a growth mindset. It is important to gently and genuinely encourage such staff, to help them overcome their barriers, and not to be too prescriptive, which risks alienating them.

Peer review can threaten some lecturers who may mistake "critique" as "criticism" and deny or become defensive about problems. It is vital to embed a positive, non-threatening, peer review culture.

Preparation of high-quality online material can be time-consuming and seen as additional, unrewarded workload, particularly when there is a reluctance to either engage with new technology or undertake additional training. With the advent of COVID, some lecturers were also required to deliver online at short notice, contributing to change fatigue.

Research output is often perceived as more important for career prospects than effective teaching. Are staff expected to excel at both research and teaching? Finally, another barrier could be a cynical view of their administration's motivation for the change.

MOTIVATORS

It is worth understanding what might motivate colleagues to adopt better practices. Motivators may be intrinsic, where the person sees direct value in the action, or extrinsic, where they seek to comply with someone else's values, possibly their employer's.

Intrinsic motivation generally produces the better results. It encompasses belief in the benefits of the change, including benefits to students, student learning flexibility, and more effective delivery. Lasting results are possible where these can be exploited. Direct extrinsic motivation such as teaching merit awards, prospects for promotion, or compliance with institution polices, is generally the next best form. Indirect extrinsic motivation includes curriculum updates, accreditation, student evaluation, peer pressure, and classroom infrastructure.

Positive Actions to Facilitate Engagement of Colleagues

One action that managers can implement immediately is to recognise that, in their organisation's workload model, changes to delivery will need time and effort. School policies and expectations should be clear in this regard.

Collegiate approaches, especially with trusted or supportive colleagues, can be effective if they are non-threatening and supportive. These could include regular community-of-practice discussions, peer review of teaching and teaching materials, nomination of change champions, examples of good practice, and pedagogy research.

Recognition of effort and achievements through teaching awards will maintain and may even increase enthusiasm for lasting change over time. It makes good work visible and might even help a colleague's promotion case. Nominate a colleague today!

Longer-term, organisational change can facilitate engagement. Newly appointed staff, perhaps replacing individuals less likely to embrace change, can have a positive effect. Staff must feel good about embracing change.

CONCLUSIONS

The post-COVID era of adapting to the new rules and requirements of online and blended teaching can be very challenging for both students and teachers. In this chapter we provided an insight into different learning techniques applicable to a blended teaching approach, and some methods that can be effectively used in online unit delivery in university level education.

While the change to blended/online learning approach from traditional lecturing methods is challenging for many, the experiences of 2020 and 2021 have shown that there is no way back. The blended learning approach has to be adopted as a "new normal". Fortunately, there are many useful methods and tools that can be adapted to specific teaching needs in different fields and levels of education. Admitting the demand to change traditional teaching approaches and implement modern online teaching techniques appears to be a fundamental pillar for entering the new era of teaching.

REFERENCES

Aragon, S. R. (2003). Creating social presence in online environments. *New Directions for Adult and Continuing Education*, 2003(100), pp 57–68.

Bates, A. W. (2019). *Teaching in a Digital Age – Second Edition*. Vancouver, B.C.: Tony Bates Associates Ltd. Retrieved from https://pressbooks.bccampus.ca/teachinginadigitalagev2/

Biriyai, A. H., & Thomas, E. V. (2014). Online discussion forum: A tool for effective student–teacher interaction. *International Journal of Applied Science*, 1(3), pp 111–116.

Brookfield, S. D. (2015). *The Skillful Teacher: On Technique, Trust, and Responsiveness in the Classroom*. John Wiley & Sons.

Castelli, F. R., & Sarvary, M. A. (2021). Why students do not turn on their video cameras during online classes and an equitable and inclusive plan to encourage them to do so. *Academic Practice in Ecology and Evolution*, 11 (8), pp 3565–3576.

Design and Delivery for Online Teaching EDU 60014, Graduate Certificate in Learning and Teaching (Higher Education), Swinburne University of Technology, online resources available via Canvas site, accessed May 2021.

Guo, P. J., Kim, J., & Rubin, R. (2014, March). How video production affects student engagement: An empirical study of MOOC videos. In *Proceedings of the First ACM Conference on Learning @ Scale Conference* (pp 41–50).

Hackman, M. Z., & Walker, K. B. (1990). Instructional communication in the televised classroom: The effects of system design and teacher immediacy on student learning and satisfaction. *Communication Education*, 39(3), pp 196–209.

Humphries, B., & Clark, D. (2021). An examination of student preference for traditional didactic or chunking teaching strategies in an online learning environment. *Research in Learning Technology*, 29.

H5P. (2021). Create, share and reuse interactive HTML5 content in your browser. https://h5p.org/

Leong, P. (2011). Role of social presence and cognitive absorption in online learning environments. *Distance Education*, 32, pp 5–28.

Lo, C. K., & Hew, K. F. (2019). The impact of flipped classrooms on student achievement in engineering education: A meta-analysis of 10 years of research. *Journal of Engineering Education*, 108(4), pp 523–546.

Merdian, H. L., & Warrior, J. K. (2015). Effective communication between students and lecturers: Improving student-led communication in educational settings. *Psychology Teaching Review*, 21(1), pp 25–38.

Pappas, C, (2013) 3 chunking strategies that every instructional designer should know, *eLearning Industry*. Available from https://elearningindustry.com/3-chunking-strategies-that-every-instructional-designer-should-know

Petrović, J., & Pale, P. (2015). Students' perception of live lectures' inherent disadvantages. *Teaching in Higher Education*, 20(2), pp 143–157.

Yee, K. (2019) Interactive techniques. *Creative Commons BY-NC-SA*, last updated 12/11/2019.

3 Developing Video Resources in Engineering Education
Evidence-Based Principles for Effective Practice

Sarah Dart and Sam Cunningham
Queensland University of Technology, QLD, Australia

Alexander Gregg
University of Newcastle, NSW, Australia

CONTENTS

DOI: 10.1201/9781003263180-5

INTRODUCTION

As technology has progressed, and the desire for accessible and flexible learning has grown, videos have become an increasingly integral element of higher education experiences (Fyfield et al., 2019). Like other disciplines, most videos used in the engineering context are direct recordings of in-person lectures (Dart, 2020; Tisdell, 2016). This is despite more innovative approaches to video design being able to capitalise on unique features of the medium like learner control (Dart et al., 2020a). Taking advantage of these features can lead to a more effective learning environment that improves motivation, increases content relevance, and supports independent study (Fyfield et al., 2019; Kay, 2012). Unsurprisingly, these benefits typically result in engineering students perceiving improvements in their learning from using videos, resulting in measurable gains in student outcomes (Belski, 2011; Green et al., 2012; Martin, 2016; Pinder-Grover et al., 2011).

The trend towards educational video usage has been greatly accelerated by the COVID pandemic, which required educators to rapidly transition courses fully online (Dart & Gregg, 2021). However, the disruptive circumstances associated with the shift influenced many educators to employ only basic approaches to video design, such as directly recording face-to-face lectures. As we move towards a post-COVID normal that increasingly relies on high-quality online learning opportunities to build student skills, purposefully developing and implementing videos within considered learning designs will be of upmost importance. In fact, Burnett et al. (2021, p. 14) highlight that educator skills in "using digital technologies" and "e-learning" will be critical to delivering on the future capabilities of graduate engineers.

A large body of research has sought to determine how learning occurs and what impacts student engagement with educational videos (Brame, 2016; Kay, 2012). However, understanding how this may be translated into practice to improve student engagement and learning outcomes can be challenging. Consequently, this chapter aims to synthesise the evidence base by highlighting important factors for consideration when producing educational video content with a focus on engineering. The chapter goes on to discuss a practical process that educators can employ when creating videos, encompassing the stages of design, production, distribution, and evaluation.

BACKGROUND

LEARNING AND TEACHING CONTEXT FOR VIDEOS IN ENGINEERING

Globally, higher education has experienced strong growth in participation over many decades, largely propelled by increasing enrolments of students from non-traditional backgrounds (Small et al., 2021). These students often have more complex life circumstances as many manage competing responsibilities such as employment and caring alongside their university commitments (Dart & Spratt, 2021). Consequently, facilitating access to learning opportunities at a convenient time and place has become critically important to supporting this cohort. As videos can be delivered asynchronously, they have become a key mechanism for enabling the required flexibility (Tisdell, 2016).

The increase in diversity of university students' backgrounds has created challenges for educators in accommodating highly varied student needs. In engineering, one key area impacted by this issue is mathematical skill level. This is due to the trend towards fewer students completing higher-level mathematics in senior secondary school (Barrington & Evans, 2016), some universities removing prerequisite knowledge thresholds for admission to engineering degrees (Office of the Chief Scientist & Australian Mathematical Institute, 2020), and the increase in mature-age learners who often lack recent exposure to formal mathematics training (Nguyen et al., 2016). Consequently, educational approaches must facilitate the bridging of students' knowledge to the required standards (Tisdell, 2016), while still developing those students who have strong mathematical skill bases. Videos are well-suited to this situation, given that they allow students to control their viewing according to prior knowledge, interests, and confidence (Fyfield et al., 2019). Thus, videos can be designed to target specific sub-cohorts in a manner not readily possible in face-to-face environments where an educator must balance the needs of the overall class (Dart, 2022).

Funding pressures (Coaldrake & Stedman, 2017) and increasing cohort sizes (Gannaway et al., 2018) have required universities to innovate approaches to teaching and student support. Employing technology has become a key avenue for this, given that technology-based solutions can often be implemented efficiently at scale (Pardo et al., 2019). Fulfilling this premise, once videos are produced, they can be readily delivered to additional students at a negligible cost. Additionally, the experience of engaging with a video is independent of cohort size, unlike face-to-face experiences, which tend to become less satisfying as student numbers escalate (Gannaway et al., 2018).

Technology also tends to be a key method for implementing universal design for learning principles that seek to improve the accessibility of learning for all (Hitch et al., 2019). For example, captions and transcripts can be readily applied to videos (unlike face-to-face teaching). This serves to benefit not only students with hearing impairments, but also those learning in their non-native language, or learning from an educator with a strong accent (Bao, 2019; Tisdell & Loch, 2017).

TYPES OF VIDEOS SUITED TO ENGINEERING COURSES

As educational videos are produced for a wide variety of purposes, many different types have emerged. Table 3.1 summarises the main types of videos applicable to the engineering context, and provides typical reasons for why each may be used in practice.

TABLE 3.1

Main Types of Educational Videos in Engineering

Video Type	Description	Typical Reasons for Usage
Animation	Dynamic illustration of a phenomenon.	Explains a complex phenomenon that can be well-represented using moving imagery (Fyfield et al., 2019).
Classroom Recording	Direct recording of a face-to-face lecture or tutorial.	Improves accessibility of in-person learning experiences for students who cannot attend on-campus or want to review the class afterwards.
Narrated Slide Presentation	Audio narration overlaid on slides sometimes with a "talking head" inset showing the speaker's delivery.	Often used to introduce an educator, course, or assessment item, as well as to provide feedback (Di Paolo et al., 2017). Pre-recording lectures in this style enables educators to design for the online experience, including making provisions to chunk content and presenting videos interspersed amongst other activities to consolidate learning.
Worked Example	Demonstration of a problem-solving process, usually involving the writing of mathematical working narrated in real time.	Allows students to pause the video to attempt work on their own at their chosen pace, as well as to rewind or skip content according to their self-determined needs (Dart et al., 2020a).
Physical Demonstration	Demonstration of a physical process or experiment.	Enables engagement with experiments that cannot be conducted in a classroom environment due to safety or cost concerns. Providing an instructional video on how to use a piece of equipment can assist learners to perform the procedure themselves (Di Paolo et al., 2017).
Software Demonstration	Walk-through demonstration of how to use software.	Allows students to replicate the process at their own pace by pausing, rewinding, and skipping (Guo et al., 2014).
Tour or Behind the Scenes	Recording at an off-campus location such as an industrial setting.	Facilitates insight into engineering sites that are not possible to visit due to logistical or economic challenges.
Interview	Recording of a speaker answering questions or discussing a topic.	Provides access to diverse or expert speakers who would otherwise be out of reach (Fyfield et al., 2019).

EVIDENCE-BASED PRINCIPLES FOR DEVELOPING EDUCATIONAL VIDEOS

A significant amount of research has sought to identify the attributes of video design that contribute to an effective learning experience (Brame, 2016; Guo et al., 2014; Kay, 2012). Here, we discuss three major aspects, adopted from Brame (2016), that should be considered when developing educational video content. These relate to cognitive processing load management, student engagement, and encouraging active learning. Table 3.2 contains a summary of the principles and how these can be applied to engineering content, which is expanded upon below.

TABLE 3.2

Summary of Theoretically-Grounded Principles for Developing Educational Videos, Including Corresponding Examples of Implementation Tailored to the Engineering Context

Aspect	Principle	Implementation Examples
Managing Cognitive Processing Load	Reduce extraneous processing	• Emphasise key information by providing a visual cue (such as a text colour change for important terms or equations). • Write mathematical working while simultaneously discussing the process.
	Manage essential processing	• Chunk content into distinct sections. • Explicitly introduce key definitions, variables, and equations. • Verbally discuss mathematical solution processes rather than providing a text-based explanation.
	Foster generative processing	• Show an animation of a process while explaining it at the same time. • Use a conversational and enthusiastic tone. • Use personal language like "you" and "I".
Facilitating Student Engagement	Minimise video length	• Keep videos as concise as reasonably possible. • Segment lectures into shorter sections.
	Ease of access	• Select a user-friendly hosting platform to distribute videos. • Use descriptive titles and timestamps. • Provide searchable transcripts and captioning. • Highlight information students commonly search for like final answers to numerical problems.
	Scaffold content	• Introduce concepts before applying them to a problem. • Link new concepts back to prior knowledge. • Ensure that content is logically structured, with content gradually increasing in difficulty. • Identify and clarify common misconceptions. • Use narration to clearly communicate reasoning behind demonstrated processes, including alternative solution paths.
Encouraging Active Learning	Prompt use of active learning during video engagements	• Build in prompts to support student reflection on key concepts. • Instruct students to pause video and attempt problems independently before watching solutions. • Implement quizzes within videos or as a follow-up activity.
	Support knowledge transfer	• Direct students to relevant resources or further problems that support their learning. • Ensure video content is connected to assessment.

Managing Cognitive Processing Load

The cognitive theory of multimedia learning explains how students learn in an environment that combines words and pictures (Mayer, 2008). The theory assumes that the human brain processes visual and verbal information through distinct channels, and that each channel has a finite capacity (Mayer, 2008; Sweller et al., 2019). Consequently, learners are only able to process a limited amount of information through each channel at a given time. Learning requires the student to actively select information from the multimedia environment, organise that information within working memory, and then integrate the information with existing knowledge structures (Mayer, 2008). To improve learning in multimedia environments, Mayer (2008) developed a series of theoretically grounded principles, which are discussed below.

First, Mayer (2008) argues that extraneous processing can be reduced by minimising the amount of superfluous information presented, while highlighting critical material using visual and verbal cues. This helps the learner appropriately select the information they draw into their working memory. Moreover, learners benefit from complementary visual and verbal information being presented simultaneously, as it can be processed efficiently through the dual channels (Brame, 2016). In the context of engineering, writing out mathematical workings in real time while discussing the process (rather than revealing a pre-made written solution) is a valuable approach to enacting this principle (Dart et al., 2020a).

Second, essential processing can be managed by presenting content in learner-paced segments that allow the student to "fully represent" concepts in their memory before moving on (Mayer, 2008, p. 765). Chunking content into distinct sections or requiring students to actively choose to move forward in a lesson are two methods for addressing this. Learners also benefit from being pre-trained on key concepts, such as where definitions, variables, and equations are explained before being applied. By familiarising the learner with the basics, some cognitive capacity is freed up for more challenging, higher-level learning tasks.

Finally, Mayer (2008) proposed that videos foster generative processing by utilising both words and pictures to present material, such as where an animation is shown while being explained verbally. The blended modalities allow students to "learn more deeply ... [as] they build connections between a verbal representation and a pictorial representation of the same material" (Mayer, 2008, p. 766). Additionally, Mayer (2008) recommended the use of a conversational rather than formal delivery style as it encourages a "sense of social partnership" between the student and educator (Brame, 2016, p. 4). Further supporting this, students prefer that educators speak relatively quickly and enthusiastically (Dart & Gregg, 2021; Guo et al., 2014). Using a personal and engaging delivery style leads to students developing a connection with their educator (Ahmad et al., 2013; Kay & Kletskin, 2012), which motivates them to try harder to understand what the educator is communicating (Mayer, 2008).

Facilitating Student Engagement

Ensuring that students find videos engaging is crucial to driving usage, thus justifying investment in video development (Dart et al., 2020a). Student perceptions of ease of use and usefulness of videos contribute greatly to engagement decisions (Dart et al., 2020a).

Research has consistently shown that keeping videos short is critical to maintaining engagement (Brame, 2016). One of the most influential studies to quantitatively demonstrate this was Guo et al. (2014). Conducted in the massive open online course (MOOC) context, the study found a significant engagement drop occurred for videos longer than six to nine minutes (Guo et al., 2014). However, subsequent research in the higher education context has observed a more linear relationship between engagement and video length (Dart, 2020). Moreover, survey-based investigations have shown that, while video length plays a major role in engagement, it is not the most significant factor in students' decision-making (Dart & Gregg, 2021; Shoufan, 2019). Therefore, while it is important that videos are kept as concise as possible, there is limited evidence to suggest that editing videos into extremely short segments (such as less than six minutes) substantially improves engagement in the university context.

Accessibility of video content strongly influences student motivation to engage. The video platform plays a role through students' familiarity with the interface, amount of downtime, capacity to vary video quality according to internet speed, ability to view videos across different devices, and capacity to interact through playlists and commenting (Dart et al., 2020a). Therefore, selecting a user-friendly platform with extended capabilities is recommended. The ease with which students can find their desired content also contributes to perceived accessibility. Consequently, developing concise videos, using descriptive titles, timestamping key segments, employing searchable transcripts, and highlighting critical information (such as final answers to numerical problems) can be strategically employed to improve findability of specific content (Dart et al., 2020a).

Scaffolding of the learning experience supports students to progressively develop skills without becoming overwhelmed and disengaged (Dart et al., 2020a; Seethaler et al., 2020). This is achieved by regulating the level of cognitive struggle, with support steadily scaled back as confidence and capability is developed (Dart et al., 2020a). Scaffolding in video-based learning designs can be implemented by clearly introducing key concepts before they are applied, while emphasising links to prior knowledge (Seethaler et al., 2020). Additionally, content should be structured such that it gradually increases in difficulty (Dart et al., 2020a). Videos are well-suited to this as they can be developed to address gaps in prerequisite knowledge, with the self-serve nature allowing those students who already possess expertise to skip the content. Research has also reported engineering students seeking extension opportunities, by such actions as asking for more difficult problems to be covered in video-based demonstrations (Dart et al., 2020a; Martin, 2016).

Related to scaffolding is the logical sequencing of content delivered within a single video. This necessitates newly introduced concepts building upon previous content without gaps in reasoning (Seethaler et al., 2020). It is particularly relevant to demonstration videos that walk students through a process in a step-by-step manner. Here the quality of explanation offered by the instructor is critical to students recognising the connections between ideas, reinforcing the rationale for why steps are being carried out in a certain way, and correcting misconceptions (Dart & Gregg, 2021; Dart et al., 2020a). The narration supports students to self-direct their learning as they are more aware of underlying reasoning as well as common pitfalls when making self-assessments of their level of understanding (Dart et al., 2020a). Further underscoring this, the literature shows that students often ask for additional clarifications in demonstration videos; for instance, asking their educator to address alternative solution paths when solving mathematically-based examples (Dart et al., 2020a; Martin, 2016).

Encouraging Active Learning

Active learning is defined by Prince (2004, p. 223) as activities where students "do meaningful learning activities and think about what they are doing". A large body of research has shown that engaging students in active learning is strongly related to increased performance, including in engineering (Freeman et al., 2014). While traditional video lectures are often designed to be watched passively in a continuous sequence (Dart, 2020), more innovative approaches to video design can support students to employ active learning techniques during their engagements.

It is worth noting that videos that demonstrate a procedure (such as solving an example problem, or showing how to use a piece of software) naturally align with the adoption of active learning strategies as students often work to replicate the process independently (Dart et al., 2020b; Martin, 2016). Dart et al. (2020b) reported that approximately 90% of students attempted the problems contained in worked example videos embedded in three engineering courses without being explicitly provoked. Students frequently use pausing to work alongside the demonstration at their own pace, as well as rewinding and skipping to focus on specific sections relevant to their needs (Dart, 2022; Dart et al., 2020b; Martin, 2016). These behaviours can also be explicitly encouraged by using verbal prompts, such as those that instruct students to pause the video to engage in the exercise before verifying their solution (Martin, 2016).

A smaller proportion of students (estimated at about 20% in Dart (2022)) report using demonstration videos to spontaneously support attempts at further problems during self-directed study. This transfer learning strategy can be encouraged by directing students to related practice problems that consolidate learning (Dart et al., 2020b). Ensuring that video content is connected to assessment (Biggs, 1996) is also a robust method for promoting knowledge transfer from video content, especially given that previous research has shown preparing for summative assessment is a dominant motivator for student engagement with video resources (Belski, 2011; Dart, 2022; Dart et al., 2020b; Key & Paskevicius, 2015).

While demonstration videos have a clear learning objective (that is, for students to be able to solve the problems independently), other styles of video, such as theoretical lectures or interviews, can have less obvious actions for students. Therefore, students can benefit from educators employing more explicit actions that encourage active learning. A useful strategy educators can consider is reflective prompts, which can aid learners in connecting prior knowledge to newly introduced concepts (Hung et al., 2014). These prompts can be implemented within a video, such as where an educator verbally poses a question and asks students to pause the video while they consider their answer, or as a follow-up potentially embedded in the learning management system (LMS). Similarly, online quizzes can be used to strengthen students' understanding of concepts taught through a video (Baker, 2016; Szpunar et al., 2014). Szpunar et al. (2014) advocate for repeated tests throughout recorded lectures to help students accurately gauge their level of performance.

PRACTICAL PROCESS FOR IMPLEMENTING VIDEOS IN LEARNING DESIGNS

Alongside the evidence-based strategies introduced above, it is important to consider the practicalities of video development. A four-stage cyclical process for implementing educational videos within learning designs is presented in Figure 3.1, broadly comprising design, produce, distribute, and evaluate stages. Each stage is expanded upon in this section.

DESIGN

The design phase of the video development process captures planning and preparation. Investing time in this early stage greatly contributes to the creation of engaging content that promotes learning (Guo et al., 2014), while helping to mitigate challenges encountered during production.

Reflecting on why videos are being considered within an educational context helps in recognising the underpinning objective and, in turn, developing a purpose-built solution that is appropriate and effective. For example, educators may recognise a need to improve access to face-to-face learning opportunities, provide students with a richer understanding of a concept, or support students to consolidate skills during their self-directed practice. Once the objective has been identified, the style of video most suited to addressing the need can be selected (noting that those styles commonly used in engineering were summarised in Table 3.1). For example, if the purpose of a video is only to allow students flexible access to classroom teaching, then a direct recording may be appropriate. In contrast, if there are challenges for students in relating new concepts to practice, developing videos that bring the "real world" into the classroom (such as through an interview with a practicing engineer or tour of an industrial site) could address the gap.

Regardless of the objective or style of video selected, educators should seek to design content that can be reused for subsequent cohorts. Strategies for maximising longevity include not referencing the semester or year of production, purposefully

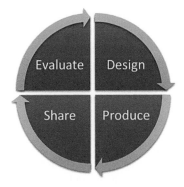

FIGURE 3.1 Process for developing educational videos.

selecting content that will remain relevant, and segmenting out sections that will ultimately need replacing (such as the introduction to a specific assessment item within a lecture).

Once video purpose and style are settled, the specifics of video content should be planned. For simple videos that can be recorded in one continuous shot, this is typically a straightforward process. For instance, with worked example demonstrations or software walkthroughs, planning may only require pre-working the solution alongside notes on key concepts to be highlighted. Similarly, to conduct an interview, preparation may only involve the development of prompting questions. These simple styles of video are often well-suited to extemporaneous delivery where the presenter speaks without notes but is familiar with what they want to say. This then readily aligns with employing the conversational approach to delivery that the research advocates (Guo et al., 2014; Mayer, 2008).

For more complex videos, such as animations or site tours, storyboarding may be considered as a planning tool (Seethaler et al., 2020). Storyboarding represents a structured method for visualising the sequence of shots, as well as the associated actions and dialogue. Scripting of the dialog may be considered in these more complex videos, especially where there is a need for highly specific wording. However, to avoid memorising the script (and thus save production time), educators may choose to record audio only when using the script, with this audio then overlaid on other imagery. Alternatively, short sections may be presented to camera, which are then intertwined around other content.

Ensuring that video resources are accessible to all students is another aspect to consider in the design phase. Web content accessibility guidelines (Web Accessibility Initiative, 2021) set out technical standards for how this can be achieved in an online environment. When applying these standards to video content, key considerations include:

- Using contrasting colours so that text and graphics are easily distinguishable.
- Selecting fonts for readability while ensuring text is large enough for viewers to engage with.
- Avoiding rapidly flashing content to minimise issues for viewers with seizure disorders.

Time should be dedicated to identifying how the evidence-based principles discussed in the previous section may be implemented within one's own context to create a positive learning environment. Table 3.2 offered a series of suggestions on how to enact the various principles within video designs related to managing cognitive loads, promoting engagement, and encouraging adoption of active learning strategies.

PRODUCE

Following the design phase, the type of recordings required for a video (such as screen capture, close-up filming of a presenter, live on-location shots, and background footage) should be clear. This then informs selection of equipment, as well as the software used for recording and editing.

Video and Audio Equipment

Many options exist for recording live shots. Portable equipment options such as mobile phones, tablets, and dedicated cameras are well-suited to filming on location, while webcams tend to be convenient when recording a "talking head" overlaid on slides or for interviews conducted via videoconferencing platforms. The required capture quality should be considered as part of the equipment selection process. For example, where the presenter's delivery is captured for picture-in-picture purposes, low-resolution built-in webcams are likely sufficient. In contrast, if recording the creation of a small electrical circuit to help students reproduce the design themselves, it is important to have high-quality video for the intricate details to be visible. In this case, a separate dedicated camera may be considered. Irrespective of the recording equipment used, the following techniques can improve filming quality:

- When capturing a presenter, ensure that the camera is positioned at, or slightly above, eye level. Where possible, the presenter should look at the camera rather than a screen.
- Ensure the subject is well lit, noting that this may also be adjusted using camera settings.
- Ensure that the background is clean and not distracting. Where this is challenging to implement, blurring the background may be considered by adjusting the camera focus or through post-production editing.

Capturing good-quality audio is important, and in many cases more important than the video quality (Shoufan, 2019). Cameras and laptops often contain built-in microphones, but these can be inadequate, resulting in recordings that are too quiet or contain distracting background noise. The latter can be particularly problematic when filming on location. Although audio can be adjusted in editing software (including reducing ambient noise), it may be necessary to use a separate purpose-built microphone to capture audio of sufficient quality.

Recording Software

Phones, tablets, and laptops often contain built-in software for filming. This software tends to be relatively simple and, as such, lacking extended functionality (such as being able to pause and resume a recording multiple times within the same file, or film several subjects simultaneously). Consequently, educators should consider more advanced software where there is an advanced need. It is worth noting that web-conferencing platforms like Zoom and Microsoft Teams are particularly useful for recording interview situations where presenters are in different physical locations (Correia et al., 2020).

Screen recording is often used to produce videos in engineering, such as when lecture slides are narrated, worked examples are demonstrated by writing on a screen, or coding is performed. Numerous software options are available for screen recording, with varying capability and user-friendliness. Those commonly available include Microsoft PowerPoint (which is well-suited to recording narrated slides), Zoom, and Microsoft Teams. Educators can also consider free and low-cost software such as Screencast-o-Matic (Dart et al., 2020b) and OBS Studio (Giles & Willerth, 2021), as well as licensed options supported by their university.

Editing

Video editing provides an opportunity to remove mistakes, join recorded clips together, and apply additional effects (like removal of background noise, addition of title cards, and creating picture-in-picture layouts). While some recording software also allows editing, functionality is often limited to basic tasks such as trimming. Similarly, although many devices have built-in editing software (such as Video Editor and iMovie on Windows and Mac devices, respectively), these tend to be quite restrictive. Consequently, where more advanced functionality is sought, options like Final Cut Pro and Adobe Premiere Pro can be considered, noting that these tend to have a steeper learning curve and costly licences. Therefore, it is worthwhile for educators to investigate what software is supported by their university.

DISSEMINATE

Methods for hosting and sharing videos should be considered in the dissemination phase of the video development process. Numerous platforms host video content, with each having its own benefits and drawbacks. Those more commonly used include YouTube (Dart et al., 2020b) and Kaltura (Giles & Willerth, 2021). Key factors to reflect upon when selecting a platform include:

- University support, such as where a platform is integrated into the university's LMS or mandated by policy.
- Capacity for learners to control their viewing experience, including pausing and resuming videos, using thumbnail images to pinpoint a specific segment of interest, and changing the playback speed (Lang et al., 2020).
- User-friendliness of engagement, including system reliability, compatibility with various devices, provisions for engaging on slower internet speeds, and extended functionality like watch history, playlists, timestamping, and commenting.
- Availability of accessibility features, including automatic closed captioning and transcript hosting.
- Capacity to integrate elements that encourage active learning, such as embedding a quiz question within a video (Baker, 2016).
- Trade-off between ease of access and privacy, given publicly accessible videos may provide the lowest barrier to entry, but can have intellectual property implications.
- Cost, given that some platforms charge to host content while others may run advertisements within videos to offset free hosting.
- Analytics capabilities that support monitoring and evaluation of video content (discussed further below).

Options for sharing videos with students can depend on the choice of hosting platform. Most simply, videos can be shared directly via a link. However, this is typically not desirable as students are taken away from the LMS. Instead, embedding videos into relevant pages in the LMS can be considered. This allows videos to appear in context, such as where text is used to introduce a video and its relevance to other content. Moreover, follow-up activities can then be coupled to the video, thus encouraging active learning in line with the evidence-based principles discussed previously (Baker, 2016). Videos may also be disseminated via playlists. This can be useful when a single learning event, such as a lecture, has been segmented into sections designed to be viewed successively.

EVALUATE

Evaluation is critical to the continuous improvement of all learning and teaching activities (Cunningham-Nelson et al., 2019; Spooren et al., 2013). With videos, there is the unique opportunity to thoroughly measure and monitor engagement as data is systematically collected at scale (Dart & Cunningham, in press; Guo et al., 2014). This forms a useful data source for understanding how videos are being used in an educational context, thus enabling educators to pinpoint both strengths of the approach and avenues for future improvement.

Most video hosting platforms automatically collect a wide variety of internal measures that can be used for evaluation purposes. Some common metrics are summarised in Table 3.3. Visualising this data can support educators in making sense of

---

TABLE 3.3

Common Metrics to Consider When Evaluating Video Implementations and Examples of How These can be Acted Upon

Metric Name	Metric Description	Example of Use in Educational Practice
View count	The total number of video views across a chosen timeframe.	Can give a high-level indication of the most popular and least popular videos, which may be related to the topics that students find challenging.
Watch time	The total length of time a video has been watched. For example, if a one-minute video was viewed twice in full, the total watch time would be two minutes.	Can provide high-level measure of student engagement, especially when normalised against view count or cohort size.
Date of viewing data	Measures when videos are accessed over a semester (an example visualisation of this data is shown in Figure 3.2a).	Can support understanding of when videos are utilised during a semester. For example, if they are being accessed weekly as part of ongoing self-directed study or mostly to prepare for assessments.
Video retention	The average percentage of a video that viewers watch (an example visualisation of this data is shown in Figure 3.2b).	Can be used to identify the specific sections students are interested in or frequently skip (Kim et al., 2014).
Time of viewing data	Measures when videos are used throughout a week on average (an example visualisation of this data is shown in Figure 3.2c).	May be used to target a particular video release time.
Individual viewer data	Data that shows the videos an individual has viewed.	May be used to prompt disengaged students towards support, or connected to other individual data (such as grades) to gauge influence of video engagement.

the information. Examples of three visualisations from the YouTube platform are presented in Figure 3.2.

In order to comprehensively evaluate the impact of videos, data collected through hosting platforms should be considered in combination with other sources (Smith, 2008). Student feedback that provides qualitative commentary on video designs is particularly useful in informing and guiding improvements. This can be obtained through surveys (such as institutional surveys performed at the end-of-semester or customised surveys designed to probe aspects of interest) or focus groups that allow for richer and more targeted feedback (Dart et al., 2020a; Dart & Cunningham, in press). Learning outcome data should also form part of a comprehensive review. However, this can be challenging to link with video engagement when the hosting platform is not integrated with institutional systems (Dart & Cunningham, in press). Participating in a thorough evidence-based evaluation process allows educators to identify improvement actions, such as recreating specific video resources to explain

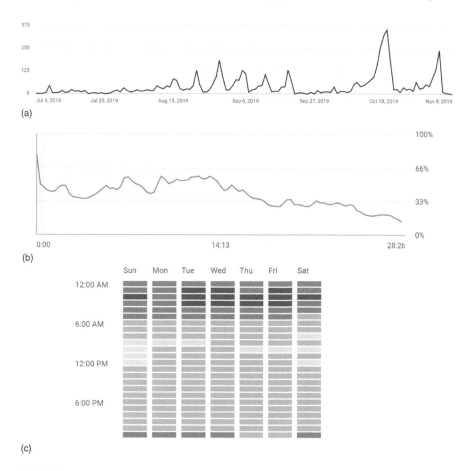

FIGURE 3.2 (a) Daily video view count over a semester, (b) average viewer retention over duration of a video; (c) heat map showing when videos are most frequently viewed throughout the week.

a concept more fully, better aligning content to students' needs when preparing for assessment, or developing additional videos where there are gaps. This then initiates the next cycle of the development process, as pictured in Figure 3.1.

CONCLUSIONS

Even as COVID disruptions in the higher education sector ease, allowing students to return to campus, videos will continue to be a fundamental aspect of learning experiences (Burnett et al., 2021). This chapter has presented an evidence-based process for developing educational videos that effectively promote learning within the engineering context. To ensure high-quality videos in a post-COVID world, it is important that educators employ these evidence-based principles across the design, produce, disseminate, and evaluate stages of video development. Educators who invest time in this process can expect to produce increasingly impactful videos as their capabilities and understanding of student learning within their context grow, ultimately leading to improved student engagement and outcomes.

REFERENCES

Ahmad, T. B. T., Doheny, F., Harding, N., & Faherty, S. (2013). How instructor-developed screencasts benefit college students' learning of maths: Insights from an Irish case study. *The Malaysian Online Journal of Educational Technology*, *1*(4), 12–25. http://research. thea.ie/handle/20.500.12065/2888

Baker, A. (2016). Active learning with interactive videos: Creating student-guided learning materials. *Journal of Library & Information Services in Distance Learning*, *10*(3–4), 79–87. https://doi.org/10.1080/1533290X.2016.1206776

Bao, A. (2019). Enhancing learning effectiveness by implementing screencasts into civil engineering classroom with deaf students. *Advances in Engineering Education*, *7*(3), 1–12. https://eric.ed.gov/?id=EJ1236925

Barrington, F., & Evans, M. (2016). *Year 12 mathematics participation in Australia – The last ten years*. Retrieved from https://amsi.org.au/?publications=participation-in-year-12-mathematics-2006-2016

Belski, I. (2011). *Dynamic and static worked examples in student learning*. Paper presented at the *22nd Annual Conference of the Australasian Association for Engineering Education*, Fremantle, Western Australia, December 5–7. https://aaee.net.au/wp-content/uploads/2018/10/AAEE2011-Belski-Dynamic_and_static_worked_examples_in_student_learning.pdf

Biggs, J. (1996). Enhancing teaching through constructive alignment. *Higher Education*, *32*(3), 347–364.

Brame, C. J. (2016). Effective educational videos: Principles and guidelines for maximizing student learning from video content. *CBE – Life SciencesEducation*, *15*(4), es6. https://doi.org/10.1187/cbe.16-03-0125

Burnett, I., Crosthwaite, C., Foley, B., Hargreaves, D., King, R., Lamborn, J., Lawrence, R., Reidsema, C., Symes, M., & Wilson, J. (2021). *Engineering change: The future of engineering education in Australia*. Retrieved from https://www.aced.edu.au/downloads/2021%20Engineering%20Change%20-%20The%20future%20of%20engineering%20education%20in%20Australia.pdf

Coaldrake, P., & Stedman, L. (2017). *Going for the higher fruit: Universities post peak public funding*. Retrieved from https://cms.qut.edu.au/__data/assets/pdf_file/0008/645029/universities-post-peak-public-funding.pdf

Correia, A.-P., Liu, C., & Xu, F. (2020). Evaluating videoconferencing systems for the quality of the educational experience. *Distance Education, 41*(4), 429–452. https://doi.org/10.1080/01587919.2020.1821607

Cunningham-Nelson, S., Baktashmotlagh, M., & Boles, W. (2019). Visualizing student opinion through text analysis. *IEEE Transactions on Education, 62*(4), 305–311. https://doi.org/10.1109/TE.2019.2924385

Dart, S. (2020). *Khan-style video engagement in undergraduate engineering: Influence of video duration, content type and course.* Paper presented at the *31st Australasian Association for Engineering Education Annual Conference*, Sydney, December 6–9.

Dart, S. (2022). Evaluating the impact of worked example videos on a large-enrolment business statistics course. *21*(1), 1–18. *Statistics Education Research Journal.* https://doi.org/10.52041/serj.v21i1.93

Dart, S., & Cunningham, S. (in press). Using institutional data to drive quality, improvement and innovation. In M. D. Sankey, H. Huijser, & R. Fitzgerald (Eds.), *Technology Enhanced Learning and the Virtual University.* Springer.

Dart, S., & Gregg, A. (2021). *Know your stuff, show enthusiasm, keep it on message: Factors influencing video engagement in two mechanical engineering courses.* Paper presented at the *Research in Engineering Education Symposium & Australasian Association for Engineering Education Conference*, Perth, Australia, December 5–8. https://aaee.net.au/wp-content/uploads/2021/11/REES_AAEE_2021_paper_226.pdf

Dart, S., & Spratt, B. (2021). Personalised emails in first-year mathematics: Exploring a scalable strategy for improving student experiences and outcomes. *Student Success, 12*(1), 1–12. https://doi.org/10.5204/ssj.1543

Dart, S., Cunningham-Nelson, S., & Dawes, L. (2020a). Understanding student perceptions of worked example videos through the technology acceptance model. *Computer Applications in Engineering Education, 28*(5), 1278–1290. https://doi.org/10.1002/cae.22301

Dart, S., Pickering, E., & Dawes, L. (2020b). Worked example videos for blended learning in undergraduate engineering. *Advances in Engineering Education, 8*(2), 1–22. https://advances.asee.org/worked-example-videos-for-blended-learning-in-undergraduate-engineering/

Di Paolo, T., Wakefield, J. S., Mills, L. A., & Baker, L. (2017). Lights, camera, action: Facilitating the design and production of effective instructional videos. *TechTrends, 61*(5), 452–460. https://doi.org/10.1007/s11528-017-0206-0

Freeman, S., Eddy, S. L., McDonough, M., Smith, M. K., Okoroafor, N., Jordt, H., & Wenderoth, M. P. (2014). Active learning increases student performance in science, engineering, and mathematics. *Proceedings of the National Academy of Sciences, 111*(23), 8410–8415. https://doi.org/10.1073/pnas.1319030111

Fyfield, M., Henderson, M., Heinrich, E., & Redmond, P. (2019). Videos in higher education: Making the most of a good thing. *Australasian Journal of Educational Technology, 35*(5), 1–7. https://doi.org/10.14742/ajet.5930

Gannaway, D., Green, T., & Mertova, P. (2018). So how big is big? Investigating the impact of class size on ratings in student evaluation. *Assessment & Evaluation in Higher Education, 43*(2), 175–184. https://doi.org/10.1080/02602938.2017.1317327

Giles, J. W., & Willerth, S. M. (2021). Strategies for delivering online biomedical engineering electives during the COVID-19 pandemic. *Biomedical Engineering Education, 1*(1), 115–120. https://doi.org/10.1007/s43683-020-00023-y

Green, K. R., Pinder-Grover, T., & Millunchick, J. M. (2012). Impact of screencast technology: Connecting the perception of usefulness and the reality of performance. *Journal of Engineering Education, 101*(4), 717–737. https://doi.org/10.1002/j.2168-9830.2012.tb01126.x

Guo, P. J., Kim, J., & Rubin, R. (2014). *How video production affects student engagement: An empirical study of MOOC videos.* Paper presented at the *1st ACM conference on Learning @ Scale*, Atlanta, Georgia, March 4–5. https://dl.acm.org/doi/10.1145/2556325.2566239

Hitch, D., Brown, P., Macfarlane, S., Watson, J., Dracup, M., & Anderson, K. (2019). The transition to higher education: Applying Universal Design for Learning to support student success. In S. Bracken, & K. Novak (Eds.), *Transforming Higher Education through Universal Design for Learning: An International Perspective*. London: Routledge.

Hung, I. C., Yang, X.-J., Fang, W.-C., Hwang, G.-J., & Chen, N.-S. (2014). A context-aware video prompt approach to improving students' in-field reflection levels. *Computers & Education, 70*, 80–91. https://doi.org/10.1016/j.compedu.2013.08.007

Kay, R. H. (2012). Exploring the use of video podcasts in education: A comprehensive review of the literature. *Computers in Human Behavior, 28*(3), 820–831. https://doi.org/10.1016/j.chb.2012.01.011

Kay, R., & Kletskin, I. (2012). Evaluating the use of problem-based video podcasts to teach mathematics in higher education. *Computers & Education, 59*(2), 619–627. https://doi.org/10.1016/j.compedu.2012.03.007

Key, J., & Paskevicius, M. (2015). Investigation of video tutorial effectiveness and student use for general chemistry laboratories. *Journal of Applied Learning Technology, 5*(4), 14–21. https://viurrspace.ca/handle/10613/2801

Kim, J., Guo, P. J., Seaton, D. T., Mitros, P., Gajos, K. Z., & Miller, R. C. (2014). *Understanding in-video dropouts and interaction peaks in online lecture videos*. Paper presented at the *1st ACM conference on Learning @ Scale*, Atlanta, Georgia, March 4–5.

Lang, D., Chen, G., Mirzaei, K., & Paepcke, A. (2020). *Is faster better? A study of video playback speed*. Paper presented at the *Proceedings of the Tenth International Conference on Learning Analytics & Knowledge*, March 23–27.

Martin, P. A. (2016). Tutorial video use by senior undergraduate electrical engineering students. *Australasian Journal of Engineering Education, 21*(1), 39–47. https://doi.org/10.1080/22054952.2016.1259027

Mayer, R. E. (2008). Applying the science of learning: Evidence-based principles for the design of multimedia instruction. *American Psychologist, 63*(8), 760–769.

Nguyen, T. H., Charity, I., & Robson, A. (2016). Students' perceptions of computer-based learning environments, their attitude towards business statistics, and their academic achievement: Implications from a UK university. *Studies in Higher Education, 41*(4), 734–755. https://doi.org/10.1080/03075079.2014.950562

Office of the Chief Scientist, & Australian Mathematical Institute. (2020). *Mapping University Prerequisites in Australia*. Retrieved from https://www.chiefscientist.gov.au/sites/default/files/2020-09/mapping_university_prerequisites_in_australia.pdf

Pardo, A., Jovanovic, J., Dawson, S., Gašević, D., & Mirriahi, N. (2019). Using learning analytics to scale the provision of personalised feedback. *British Journal of Educational Technology, 50*(1), 128–138. https://doi.org/10.1111/bjet.12592

Pinder-Grover, T., Green, K. R., & Millunchick, J. M. (2011). The efficacy of screencasts to address the diverse academic needs of students in a large lecture course. *Advances in Engineering Education, 2*(3), 1–28. https://eric.ed.gov/?id=EJ1076056

Prince, M. (2004). Does active learning work? A review of the research. *Journal of Engineering Education, 93*(3), 223–231. https://doi.org/10.1002/j.2168-9830.2004.tb00809.x

Seethaler, S., Burgasser, A. J., Bussey, T. J., Eggers, J., Lo, S. M., Rabin, J. M., Stevens, L., & Weizman, H. (2020). A research-based checklist for development and critique of STEM instructional videos. *Journal of College Science Teaching, 50*(1), 21–27. https://www.nsta.org/journal-college-science-teaching/journal-college-science-teaching-septemberoctober-2020/research

Shoufan, A. (2019). Estimating the cognitive value of YouTube's educational videos: A learning analytics approach. *Computers in Human Behavior, 92*(2019), 450–458. https://doi.org/10.1016/j.chb.2018.03.036

Small, L., McPhail, R., & Shaw, A. (2021). Graduate employability: The higher education landscape in Australia. *Higher Education Research & Development*, 1–15. https://doi.org/10.1080/07294360.2021.1877623

Smith, C. (2008). Building effectiveness in teaching through targeted evaluation and response: Connecting evaluation to teaching improvement in higher education. *Assessment & Evaluation in Higher Education, 33*(5), 517–533. https://doi.org/10.1080/02602930701698942

Spooren, P., Brockx, B., & Mortelmans, D. (2013). On the validity of student evaluation of teaching: The state of the art. *Review of Educational Research, 83*(4), 598–642. https://doi.org/10.3102/0034654313496870

Sweller, J., van Merriënboer, J. J., & Paas, F. (2019). Cognitive architecture and instructional design: 20 years later. *Educational Psychology Review*, 1–32. https://doi.org/10.1007/s10648-019-09465-5

Szpunar, K. K., Jing, H. G., & Schacter, D. L. (2014). Overcoming overconfidence in learning from video-recorded lectures: Implications of interpolated testing for online education. *Journal of Applied Research in Memory and Cognition, 3*(3), 161–164. https://doi.org/10.1016/j.jarmac.2014.02.001

Tisdell, C. C. (2016). *How do Australasian students engage with instructional YouTube videos? An engineering mathematics case study.* Paper presented at the *Australasian Association for Engineering Education Annual Conference*, Coffs Harbour, Australia. https://aaee.net.au/wp-content/uploads/2018/10/AAEE2016-Tisdell-Instruction_YouTube_video_Australian_student_engagement.pdf

Tisdell, C., & Loch, B. (2017). How useful are closed captions for learning mathematics via online video? *International Journal of Mathematical Education in Science and Technology, 48*(2), 229–243. https://doi.org/10.1080/0020739X.2016.1238518

Web Accessibility Initiative. (2021). Web Content Accessibility Guidelines (WCAG) Overview. https://www.w3.org/WAI/standards-guidelines/wcag/

4 The Effective Learning Strategies for Teaching in COVID and Beyond

Connecting and Aligning Content with Context

Christopher Love and Julie Crough
Griffith University, QLD, Australia

CONTENTS

INTRODUCTION

No one could have imagined the transformation that occurred in higher education due to the COVID-19 pandemic. The shift from traditional lecturing on campus to remote teaching online literally happened overnight, with universities and academics scrambling to adjust to the technology of online learning platforms for class delivery. In addition, many academics were learning to teach online for the first time, grappling with an array of questions about what were the best practices for student engagement in the online space? At the time, there was little time to redesign and implement best practices, although academics did their best under the circumstances. It is interesting to reflect on some of the first online classes we delivered and our frustrations with the online learning platforms, as well as the issues we encountered with the internet. Students stuck in break-out rooms, classes where not one student had a working microphone, cut and paste from Wikipedia in online exams, students logged into online classes but clearly 'not there', were just a few of the frustrations, and while not funny at the time, upon reflection, these antics have provided some humour. On a more positive note, in order to adapt to the disruption, many courses were redesigned to improve remote learning and support students to learn more independently, and these practices remain in use post-COVID. Certainly, traditional lectures may well be a thing of the past, as blended learning approaches appear to be the future direction of higher education. According to the 2021 Horizon Report, "learning models that enable flexible movement between remote and in-person experiences will help institutions minimise disruption and ensure continuity of course delivery through future crises" (Pelletier et al., 2021, p. 8). Post-COVID teaching in face-to-face or online classes in the future will be focused on a variety of active learning tasks aligned with assessment, designed to apply knowledge provided in pre-recorded lectures. Research has demonstrated that active learning strategies are effective in reducing the achievement gaps for underrepresented students in undergraduate STEM courses (Freeman et al., 2014). More recently, Theobald et al. (2020, p. 6476) found that, while active learning benefits all students in STEM-related courses, it "offers disproportionate benefits for individuals from underrepresented groups".

In this chapter, we will outline the practices that we introduced into a large first-year biochemistry course to support student learning prior to and during COVID-19, outlining the reasons for maintaining these strategies post-COVID, as well as discussing how these strategies could be adopted in similar engineering courses that are content-heavy, require students to master threshold concepts, and are cognitively challenging.

CHALLENGES OF THE FIRST YEAR – A PERSPECTIVE FROM BIOCHEMISTRY

Fundamentals of Biochemistry is a large first-year course with 300–400 students at a research-intensive university in Brisbane. The course is high in content with many threshold concepts for students to navigate, and as a result, had historically high rates of failure in 2017 and 2018 (37% and 38%, respectively). In 2019, an inspection of the student demographics and ethnicity revealed that the course had a very diverse cohort, with a high percentage of students who were first in family (>45%, 2019), in addition to students for whom English was their second language (>30%, 2019), and those from low socio-economic backgrounds (>21%, 2019). In addition, the entry requirements varied significantly, ranging from students enrolled in the Bachelor of Science with low university requirements to those with direct entry into medicine, who are required to have the highest university entry requirements. This results in students having a wide variety of abilities, and we pondered over whether the low-entry students had the requisite skills to transition through university. Such students, "with lower levels of academic preparedness, critical differences in social and cultural capital, and often ill-formed expectations of what to expect and what is expected of them" (Kift, 2015, p. 52), then faced further challenges in the academically demanding STEM disciplines (Freeman et al., 2014). Interestingly, this is not a new issue, as researchers advocated two decades ago that "universities need to make realistic appraisals of the academic demands of particular courses and set the entry score for school leavers at a level based on academic challenge rather than on course demand" (McKenzie & Schweitzer, 2001, p. 31).

Understanding the diverse cohort of students in the course and catering for their needs triggered an initial restructure of the course in 2019 to improve the student success rates. The assessment was redesigned for learning and students were required to develop personal study plans, incorporating learning strategies, which were reported to enhance performance (Kift, 2009; Dunlosky et al., 2013; Elkington, 2015). A decrease in the fail rates from 38% to 27% was observed after the first offering of the redesigned course, suggesting these strategies improved student success.

The COVID-19 pandemic forced further changes in the course to support students learning remotely. Based on our initial experience of teaching online, remote learning requires students to be more disciplined, and more responsibility is placed on students to manage their studies. In addition, students in the first year have been off-campus since the start of their university experience and have not had opportunities to befriend their fellow classmates, form support networks or study groups, or feel part of the university community. We countered these issues by incorporating assessment into a comprehensive scaffolded digital learning journal, segregating the content into weekly chunks, aligning the learning journal activities with problem-based online tutorials as well as embedding the assignments, and student evaluations and reflections. Moreover, a Microsoft Teams site was created to build a sense of community within the course. The "biochemistry community" contained channels for students to interact socially to meet their colleagues or discuss biochemistry. Although challenging, the changes introduced saw further improvements in student success (including reduced fail rates to 20–21%) through the COVID years of 2020 and 2021.

PRE-COVID COURSE IMPROVEMENTS

The first steps to improve first-year biochemistry occurred in 2019, prior to the COVID-19 pandemic, to deal with the very high fail rates. We introduced two changes: (1) provide multiple attempts for online quizzes to support learning rather than penalise students for a single poor quiz performance; and (2) ensure that students gained the necessary skills for success at university by having them develop personal study plans, incorporating learning strategies that improve performance.

ONLINE QUIZZES – MULTIPLE ATTEMPTS TO SUPPORT LEARNING

Many students are under-prepared for the course assessment, particularly early in the trimester with courses that are heavy in content and contain many threshold concepts. This leads to poor outcomes for many students, and they are inadvertently penalised for a single poor quiz result. Our strategy was to introduce multiple attempts of each quiz over a defined time-period. This provided students with an opportunity to complete the quiz, reflect on their performance, identify their weaknesses and revise using their learning resources, and then take the quiz again, hopefully improving their result. The majority of students did improve their quiz result as an outcome of this strategy, and we received many student comments in this regard. For example, *"Being able to take the quiz multiple times was fantastic. This allowed me to revisit the content after my first attempt to see what I might be missing, and gave me the opportunity to improve my results"*, and *"The three attempts were also valuable as you got to see where you went wrong and to study a little bit more and go back to try again"* (student comments, 2019). If quizzes were a single attempt, an additional 19 students (7.4%) would have failed the first quiz and 23 students (9.0%) the second. Therefore, this strategy has supported student learning and contributed to student success in this course.

PERSONAL STUDY PLANS – PERFORMANCE EVALUATION, REFLECTION AND CREATION OF STUDY PLANS

The transition from secondary to tertiary education can be more challenging for traditionally underrepresented students who are often ill-prepared for the demands of STEM-based courses (Freeman et al., 2014). With this in mind, and a student cohort in first-year biochemistry with a high percentage of the class being first in family, speak English as a second language, or come from low socio-economic backgrounds, it begs the question, *Are these students entering university with the necessary skills to successfully transition through their degree programmes?* If not, then we need to ensure that we are developing these skills and exposing students to a range of strategies to support their learning. Research across four decades has demonstrated that study skills and study strategies can influence academic performance (Pantages & Creedon, 1975; Dunlosky et al., 2013; Hora & Oleson, 2017; Sebesta & Speth, 2017).

Our strategy involved students creating a personal study plan to improve their performance in online quizzes and the final examination. However, prior to study plan creation, students were required to evaluate and provide a reflection of their quiz performance to develop their ability to become self-regulated learners (Butler, 1998; Boekaerts et al., 2000; Zimmerman, 2000). There are a multitude of benefits to student learning associated with self-assessment and reflection activities, which have been recently reviewed by Chan and Lee (2021). Reflections designed to be critical of performance can guide students to be self-aware, identify their strengths and weaknesses, encourage them to take greater ownership and responsibility for their learning, and support students to become self-learners (Mann et al., 2009; Bond et al., 2011; Tsingos et al., 2014; Chan & Lee, 2021).

Following evaluation and reflection of their performance in the first online quiz, students were required to develop a study plan to maintain or improve their grade from the first quiz to the second. To develop their study plans, students were provided with a simplified version of learning strategies (see Table 4.1) that were shown to improve performance, and they were required to incorporate two of these learning strategies in their study plans (Dunlosky et al., 2013; Hora & Oleson, 2017; Sebesta & Speth, 2017). This process, evaluation of quiz performance, and creation of a study plan, was repeated after the second online quiz (mid-trimester) to encourage students to prepare for the final examination. Two of the simplified learning strategies were again required, although it was suggested that students consider which strategies would be appropriate for the final exam, which covered all content in the course and contained a combination of short answer and multiple-choice questions compared to the online quizzes, which contained only multiple-choice questions. Students were

TABLE 4.1

A List of Simplified Learning Strategies that have Shown to Improve Learning and Understanding in Science, Technology, Engineering, and Mathematics (STEM) Disciplines

Learning Strategies	Methods of Applying these Learning Strategies
Self-evaluation	• Check the progress of your own work • Monitor understanding of material • Address or clarify confusions or gaps in knowledge • Complete tutorial questions prior to class and compare your answers • Reflect on your own understanding of concepts
Reviewing graded work	• Review your own graded work (quizzes, assignments, etc.) • Perform practice quizzes and reflect on performance • Correct assignment errors to correct misunderstandings
Seeking information	• Attending Peer-Assisted Study Sessions for revision and reinforcement of concepts • Supplement lecture notes with outside resources not provided in class (YouTube videos, websites, alternative textbooks) • Seek help and ask questions (other sources outside the class) • Study with friends and seek information from peers
Keeping records and monitoring	• Write or type notes when studying from a book, lectures, etc. (in class and outside the class) • Mark content that you don't understand • Rearrange material/information into a format that makes learning more effective (bind own notes, create chapter outlines and a study guide, develop note cards)
Goal setting, planning, and time management	• Make a timeline to prioritise study tasks and materials • Begin studying earlier and in advance of the exam • Keep up with assigned readings • Allocate specific times to study • Use study time effectively
Reviewing lecture notes and textbook resources	• Review class notes, instructor notes, lecture slides, and textbook readings • Organise lecture notes, summarise lectures and topics • Review tutorial questions

Dunlosky et al. (2013), Hora and Oleson (2017), and Sebesta and Speth (2017).

also required to evaluate and reflect on developing personal study plans, indicate whether they had developed a study plan previously, and note whether the study plans better prepared them for the assessment (Likert scale question). The student evaluation and reflections provided a direct insight into students' perceptions of the success of creating study plans, which we used to determine the effectiveness of this strategy (Love, Crough, Green, & Allan, 2020).

The study plans produced by students were both creative and elaborate (see Figures 4.1 and 4.2), and the results from a student evaluation with a response rate of 95% suggested that implementing personal learning plans and reflection was successful. Students reported that they maintained (41.1%) or improved (30.2%) their grades on Quiz 2 compared to Quiz 1. This was encouraging, considering that Quiz 2 covered a larger volume of content and arguably more difficult concepts, suggesting that the personal study plans may have played a role in the improvements in performance that were observed. In addition, 33.5% stated that their study plan contributed to improving or maintaining their grades, although this was despite only 7.7% of students reporting that they fully enacted their study plans. The majority of the students commented positively in relations to the study plans. Examples from student reflections (2019):

> Constructing a detailed study plan using the six study strategies has been vital in my quiz results this trimester. It has helped me manage my study time more effectively with certain goals in mind.

> Developing a study plan supported my learning and resulted in improved grades. By giving me a clear and easy applicable plan, I was able to better focus and direct my study efforts in a more productive, efficient, and controlled manner.

Overall, 72.4% of students indicated that developing and using a study plan was useful (52.3%) or very useful (20.1%) in preparing them for the assessment in the course. However, this indicates that 27.6% of students did not find creating a personal study plan useful.

Some students indicated that the study plans were not helpful because they were unable to effectively follow them. For example, *"Developing a study plan has not helped me improve my grades in biochemistry. I found the plan impossible to stick to, personally"* (student reflection, 2019). Further investigation revealed a myriad of reasons why only 7.7% of the students fully enacted their study plans. The most common reasons were assessment for other courses (15.3%), work commitments (14.5%), poor time management (12.9%), and unidentified personal reasons (12.9%). More concerning reasons for not following study plans were sickness (8.8%), a lack of motivation (7.2%), and mental health issues (2.8%). We have tried to alleviate these issues in future offerings of the course by providing guidance to students, suggesting they take into consideration their commitments outside this course when creating a study plan and ensuring that their plans are realistic and achievable, which may enhance motivation and reduce stress. Students that followed their plans echoed these sentiments,

> I noticed that having a study plan for Fundamentals of Biochemistry meant that I had better time management and therefore less stressed throughout the trimester. This therefore impacted on my successful grades and increased my confidence for the final exam.
>
> (student reflections, 2019)

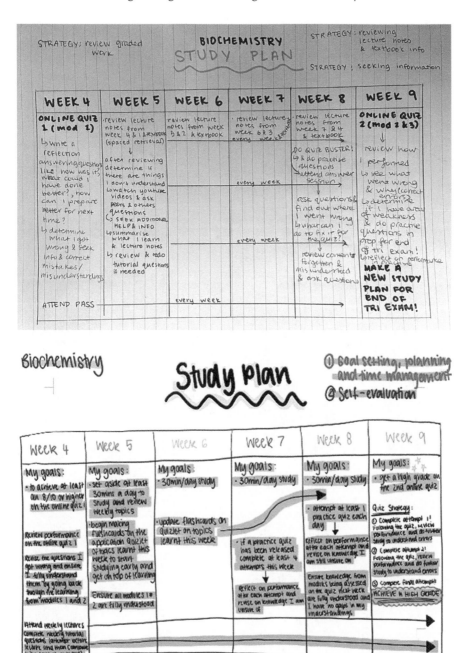

FIGURE 4.1 Examples of student study plans to prepare for the mid-trimester quiz.

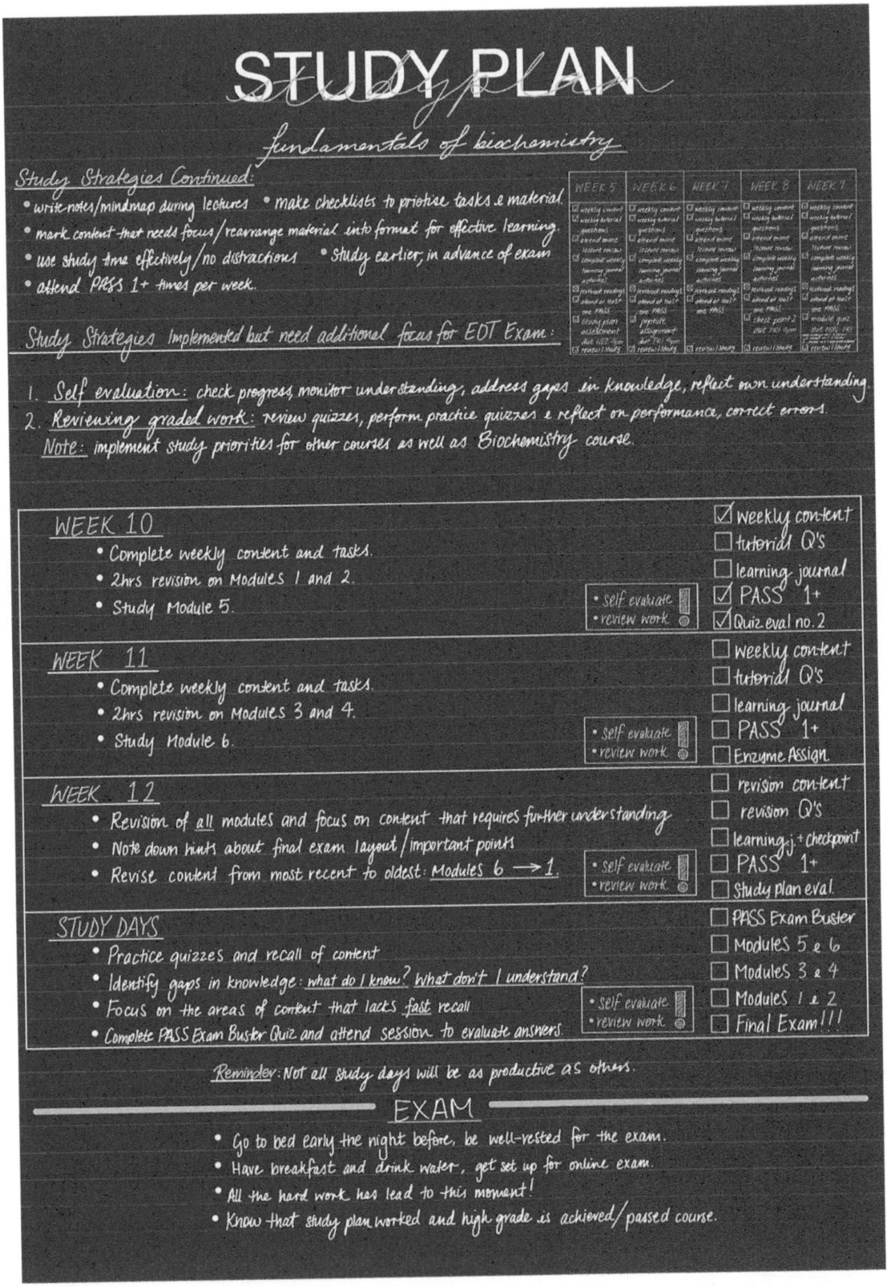

FIGURE 4.2 An example of a student study plan to prepare for the end of trimester exam.

Surprisingly, 60.9% of students indicated that they had never developed a study plan for another course prior to commencing university, which supports the notion that not all students are entering higher education with the study skills required to be successful. However, we were encouraged that 88.2% of students stated towards the end of the course that they were capable of developing study plans for future courses. Creating study plans as part of first-year biochemistry will remain post-COVID with students identifying and commenting on the benefits of this strategy,

> I feel that the study plan helped me with my learning. I found the biggest benefit was that it gave me weekly goals to work towards and broke my study down to much more manageable parts. This allowed for a more structured and efficient study regime.
>
> (student reflection, 2019)

COVID COURSE IMPROVEMENTS

As previously mentioned, with the onset of the COVID-19 pandemic there was an immediate shift to remote learning and, as staff, we had to quickly adapt to teaching online. However, first-year biochemistry is in Trimester 2, providing us with an opportunity to reflect on our online teaching experiences in Trimester 1 and improve the remote learning experience for students in future courses, using lessons learned. Our first concern was that students in the first year had only a couple of weeks on campus before going into lockdown and probably hadn't had the opportunity to meet other students in the course, form relationships, or build networks to support their learning. In addition, remote teaching relies on students taking control and being responsible for their own learning as well as managing their time across several courses. To support these challenges, we introduced additional changes in the course: (1) A Microsoft® Teams site with channels for social interactions and discussions regarding biochemistry to build a course community; (2) Dividing the content into weekly, manageable tasks aligned with problem-based online tutorials; and (3) Incorporating weekly biochemistry tasks, creation of study plans, and assignments into a single digital learning journal (ePortfolio) along with student evaluations and reflections.

BUILDING A SENSE OF COMMUNITY IN AN ONLINE LEARNING ENVIRONMENT

One of the challenges of remote learning is the lack of direct interaction with the students in the class. It is the informal conversations before and after classes that help build rapport and make connections with students. Beins (2016) has suggested that social or informal conversation are just as important to relationships and learning, as engaging academically in the course content. As a lecturer and course convenor, I initiated a community space for my second-year biochemistry class as we were thrust into COVID-19 lockdown, with avenues for students to interact socially or discuss the course content (Love, 2020). In this case Microsoft® Teams was used for social interaction with and between students and teaching staff, and photographs of my dog and pictures of the sunset from my backyard were used to initiate informal conversations to get to know my students, while an online discussion board was used for students to discuss the content relevant to the course content. Discussion boards (prior to and during the COVID pandemic) are beneficial in providing a voice for students, increasing their communication skills, and promoting knowledge extension and metacognition (Anderson, 2006; Beins, 2016; Rashtchi & Khoshnevisan, 2021). A discussion board was initiated with a thread on first structures of the Coronavirus spike protein complexed with an antibody published, a topic not only relevant to current real-world issues but also directly related to the course, which focuses on the structure and functions of proteins. Many students were actively engaged in the community site and discussion board, commenting favourably.

> The extra steps taken to ensure that the students could still communicate with each other (about both course content and engaging in discussions regarding proteins) helped with still feeling connected despite learning taking place online. Also, the opportunity for the students to be involved in some way with choosing topics and the assessment plan helped very much with feeling connected to the course content.
>
> (Student evaluation of Course, 2020)

Following the success of a community site for both social interaction and disciplinary conversations between students and academic staff in the second year, we immediately introduced this strategy into our first-year biochemistry course. In fact, we believe having a platform for student to connect with each other was crucial because our first-year students had only experienced two to three weeks on campus before the shift to remote learning. These students had little or no time to meet other students in the class, build relationships, or form support networks. Therefore, providing avenues for students to connect with each other and teaching staff was critical for enhancing engagement and learning in the class. The "Biochemistry Network" provided an opportunity to meet other students and/or discuss the course content.

Over the past two years of COVID remote teaching in this class we have seen much engagement through the "Biochemistry Network", such as students sharing personal information, forming study groups, and engaging in discussions pertaining to biochemistry. Many students posted questions on the network and, although academic staff were actively monitoring the site, many of the questions posted were answered by other students before academics could respond. The aforementioned aspects of the network are a measure of its success, even though only a portion of the class chose to use this platform. There were many comments on student evaluations of the course that were supportive of academic staff involvement, for example, *"made an effort to connect with students and be open to questions via Teams"*; *"actively engaged in the social side of learning"*; and *"encouraging discussion of biochem and nonrelated topics outside of class, especially given the isolation of distance-learning"*. While courses post-COVID will probably involve blended teaching, there will be a need to provide opportunities for students to meet and interact outside the classroom.

WEEKLY BIOCHEMISTRY PROBLEM-SOLVING

Traditionally, this course consisted of a combination of lectures and whole class tutorials. During COVID, the traditional lectures were replaced by pre-recorded lectures and online tutorials focused on problem-based learning. This remote teaching model does not necessarily support students in time management or them to become responsible for their own learning. Therefore, we decided to use a digital workbook (PebblePad), which divided the content into manageable weekly chunks, encompassing key concepts and content from the course, as well as weekly problem-solving activities (see Figure 4.3). The biochemistry problem-solving activities were aligned with the online problem-based tutorials to scaffold learning. Feedback on the problem-solving provided further scaffolding and this was provided at checkpoints throughout the trimester. The goal was to initiate independent and self-regulated learning, and support students with their time management as well as develop their own study resources which could be used for exam preparation.

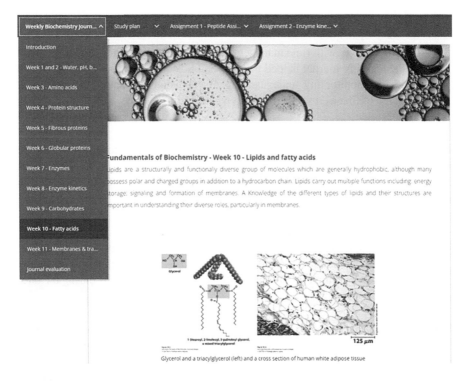

FIGURE 4.3   The biochemistry learning journal showing the weekly topics and an example of the Week 10 worksheet.

Student evaluations used to judge the learning benefits of the weekly biochemistry activities revealed that 67.7% and 67.1% of students stated that the weekly activities were effective or very effective in 2020 and 2021 (respectively), with 72.2% and 63% (respectively) indicating that the activities were aligned or very well aligned with the problem-based tutorials and lecture content. In addition, 57% and 65.4% (in 2020 and 2021, respectively) of students agreed that weekly biochemistry tasks helped them manage and keep up with the course content. The results of this evaluation suggest that the weekly activities compiled in a digital workbook supported the students in navigating and managing the course material. Here are two of many student reflections that supported this notion:

> Being able to seamlessly transition between various activities made it easier to develop connections between the course content, therefore effectively enriching my understanding of key concepts. I also found that the weekly exercises provided a great framework for weekly study, allowing me to quantify the amount of study I was accomplishing per week. The questions themselves were also very appropriate in regards to the course learning and encouraged critical thinking; thus, after completing the exercises each week I always felt very comfortable with the lecture material [...]

and:

> I tend to struggle with time management, especially with other courses. I found that the learning journal helped me stay on track with learning material. It also helped me monitor my understanding of class content, and so I really took advantage of that as my starting point when studying.
>
> (student reflections, 2020)

One of the goals of the weekly activities was to support students to develop effective time management skills, which was revealed in a student survey of the study plans introduced in 2019. Many students reflected on the weekly biochemistry activities in this regard:

> [The] biochemistry learning journal was [a] good means of reinforcing understanding of topics week-by-week as content was covered in lectures and tutorials the decreased amount of tutorial questions to focus on journal entries was good for balancing learning journal Work with other assessment items during the later weeks of the term.

The learning journal was also designed to provide students with content, concepts, and examples with students commenting favourably, *"The journal covers key concepts and equations utilised in exams and assessment as well as tutorial questions"* (student reflections, 2020). Based on the results and student reflections, it appears that the weekly biochemistry activities have supported time management and student learning.

DIGITAL BIOCHEMISTRY LEARNING JOURNAL FOR COVID-19

Finally, to support remote learning in the pandemic we designed a comprehensive biochemistry learning journal (ePortfolio) in PebblePad, which combined the weekly biochemistry activities (outline above), creation of personal study plans (introduced in 2019), and course assignments (previously paper-based). The PebblePad learning platform provided a single access point for students to complete several assessment items and provide a safe space for reflect of their performance, and despite the diverse student cohort from a range of degree programmes, most students in the biochemistry course were familiar with the PebblePad personal learning platform, having used it in their first trimester chemistry course, which was a pre-requisite for biochemistry. Our aim was for students to not only complete the tasks and assessment in the learning journal but develop this into an ePortfolio to use as a study resource for exam preparation. ePortfolios were chosen for biochemistry as they are considered high-impact educational activities (Kuh, 2008), particularly when they include reflective practices (Eynon & Gambino, 2017). Reflection on learning had already been implemented in the pre-COVID course redesign and activities but was expanded across the student digital learning journal. *"Reflective pedagogy transforms ePortfolio from a push button technology into an emerging process of connection, integration of academic learning life experience and profound processes of personal growth"* (Eynon & Gambino, 2017, p. 40). After implementing ePortfolios to scaffold and enhance students' reflection in higher education, Roberts et al. (2016) developed three key design principles. These principles informed some of the design elements of the learning journal that were developed. First, use the ePortfolio in the weekly tasks to encourage students to use the journal as part of their weekly routine (Roberts et al., 2016). The biochemistry learning journal was designed with weekly content and threshold concepts with related tasks including problem-solving to help students develop regular study habits. Second, ensure the task is clear and demonstrate to students how it is linked to their learning (Roberts et al., 2016). This design principle was achieved by aligning the learning activities with problem-based tutorials and embedded key concepts in the learning journal. Third, Roberts et al. (2016) recommend that the ePortfolio should be implemented from the outset of a student's degree, and, as mentioned, the PebblePad learning platform had been introduced in the pre-requisite chemistry course.

A survey and student reflections were used to determine the success of the biochemistry learning journal. Overall, 71.8% and 72.7% of students indicated that the learning journal was effective or very effective in supporting their learning in the course for 2020 and 2021, respectively. Students also highly rated the effectiveness of having the study plans, biochemistry activities, and assignments in one digital learning platform, with 76.8% and 72.7% of students agreeing for 2020 and 2021, respectively. Many student reflections also commented the ease of access to a single platform to complete assessment and being able to return and correct their work at a later time. For example, *"I really liked how everything was available and accessible in one space"*, and *"The biochemistry learning journal was easy to access, and the fact that it was able to be saved and available for later use was effective"*.

Other students commented on the learning journal helping them to revise the content or become more organised, such as,

> I really enjoyed using this biochemistry learning journal as it helped me revise content each week. I think it is very useful and it helps me become more organised as the assignments, biochemistry tasks, and study plans are all in the same digital workbook.

Larmar and Lodge (2014) have suggested that metacognition is a critical factor in first-year transition, and there were many examples of metacognition as well as self-regulated learning in student reflections, for instance:

> I found the biochemistry learning journal to be a unique, yet effective, learning tool throughout the trimester. I felt that it served as a good tool in terms of checking what content I understood, or didn't, from the lecture content in that week. As well, it forced me to synthesise the information I had learnt into my own words; which, in its unique way, allowed me to grasp the concepts further and identify any possible gaps in my knowledge.

However, it is worth noting that not all students were in favour of the learning journal; for example: *"At times, the learning journal seemed to be a bit of a burden as I often felt that I was spending too much time doing the questions and then uploading them to the journal"*, and *"I found it stressful to have the learning journal to complete on top of all the other assignments that I had for this course as well as my other courses"*. Students did comment that they struggled to keep up towards the end of the trimester as the content became more difficult or as assessment was due for other courses. Students were frustrated with technical difficulties associated with uploading documents, pictures, and/or photos even though instructions were provided to minimise these issues. With some minor adjustments and more informative instructions we will improve these technical issues for future course offerings. The overwhelming majority of students commented positively that the biochemistry learning journal was helpful and supported learning and improved time management, providing a valid argument for the learning journal to be an integral part of the assessment for this course in the future.

POST-COVID TEACHING

COVID dramatically changed the way we taught when remote teaching was required during pandemic lockdowns. What teaching practices remain post-COVID depends on reflecting on our current practices and understanding which strategies have improved students' learning during the pandemic. We have outlined the redesign of a large first-year course to improve student learning and ensured that students have the required skills to transition through university. COVID may have changed the way we teach but course improvement should not be governed by a pandemic. We should be regularly reflecting on our teaching practices and adjusting these to support the needs of current students; this will lead to improved course outcomes and student success. This is the approach taken for our first-year biochemistry course prior to and during the pandemic. The improved student success can be attributed to alterations made in this course since 2019, which are outlined above and shown in Figure 4.4. Beyond COVID, the mode of teaching and content delivery will be the focus to future improvements in first-year biochemistry to further improve student engagement and encourage self-regulated learning.

Academics are faced with a number of challenges associated with teaching post-COVID as we transition back to campus, particularly in large first-year courses with a diverse range of students. Students have preferences for the types of teaching they prefer: some prefer face-to-face classes for accountability, while others have a preference for online classes for inclusiveness and flexibility. Academics, on the other hand, with increasingly high workloads, struggle to accommodate all students' preferences. The key to post-COVID teaching is finding the "right" combination of face-to-face and online classes in a blended learning setting. Blended modes of teaching are an amalgamation of technology and classroom teaching strategically designed to enhance learning (Garrison & Kanuka, 2004; Torrisi-Steele & Drew, 2013). In biochemistry, our blended approach involved providing pre-recorded lectures and a set of tutorial questions to be completed prior to attending a problem-based tutorial, and the alignment of these activities with the completion of a learning journal in PebblePad (biochemistry learning journal). This is the traditional "flipped classroom" approach, although the problem-based tutorials during COVID were delivered online in 2020, and via hybrid classes in 2021, streaming to students online from a face-to-face class. Regardless of the delivery mode, the "flipped classroom" approach has it challenges, most notably, the inability to deliver an effective problem-based

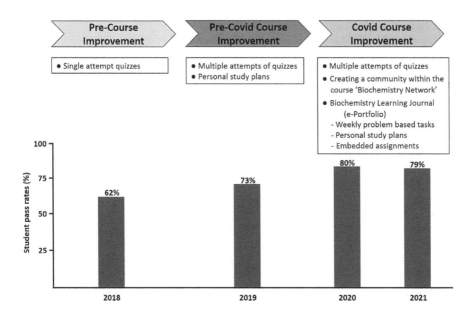

FIGURE 4.4 Course improvement strategy showing the increase in student pass rates.

tutorial if students haven't watched the pre-recorded lectures and/or completed prior learning activities. Providing access to recordings of problem-based tutorials for flexibility during COVID resulted in low attendance rates; this was observed for both online and hybrid classes. In addition, the hybrid classes were hampered by multiple technical problems including software, hardware, and sound issues as well as poor internet connection, which resulted in students logging out of the online sessions. As students transition back to campus, how do we motivate students to complete required learning activities prior to face-to-face classes to ensure an effective and engaging experience?

We have thought about making tutorials compulsory although we are not convinced that this will improve student motivation or engagement but rather increase the workload of academics required to monitor attendance. We have also considered using adaptive release of course material to ensure that students prepare for the classes, but this creates more assessment deadlines and added pressure on students who are already struggling with their own wellbeing during and post-COVID. Alternative teaching approaches are required for biochemistry post-COVID and these need to revolve around improving the blended teaching approach and content delivery, to motivate students to take control of their own learning and to ensure that problem-based classes provide an effective and authentic learning experience. The strategies we are considering for biochemistry as we transition out of COVID and back onto campus include: (1) Embedding laboratories into the course to connect theory with practice; and (2) Introducing a Students as Partners (SaP) component to empower students to take ownership of their learning as well as improve motivation and engagement in course content.

EMBEDDED LABORATORIES IN THE FIRST YEAR

When we were undergraduate students, many STEM courses contained a laboratory component, which was designed to link the theory to practical laboratory skills. Many first-year courses in STEM still contain laboratories; however, the laboratories in some courses, including first-year biochemistry, were removed many years ago to create laboratory courses. In many cases, the laboratory courses were running in parallel to courses covering the theory and the connection between theory and practical application were maintained. However, over many years of degree programme re-structures and course re-development, the connection between theory and practice is less distinguishable. For example, the laboratory experiments related to first-year biochemistry are in a second-year laboratory course. This creates several learning challenges for students. First, many students tend to create "cognitive silos" separating one course from another, and struggle to connect the theory when performing laboratory experiments. Second, when courses contain only theory, it is difficult for students to understand the reasons for learning certain concepts. We believe re-introducing laboratories into the first year provides a clear connection between theory and practice, which will support the understanding of such concepts and provide authentic learning experience for students. Laboratories will be introduced into first-year biochemistry this year (2022) and it will be interesting to determine their effectiveness on student learning of the associated key concepts.

STUDENT-STAFF PARTNERSHIPS IN TEACHING AND LEARNING

One strategy we are keen to introduce into the first year is a Students as Partners (SaP) approach, which involves providing an opportunity for students to have a voice in teaching and learning. This approach empowers students to take ownership of their learning by participating with staff to develop, co-create, and negotiate course curricula (Bovill et al., 2011; Cook-Sather et al., 2014; Healey et al., 2014). We have successfully introduced a student-staff partnership into second-year biochemistry (protein science), which increased student engagement and motivation by allowing students to vote for topics they would like to learn as well as co-create assessment (Crough & Love, 2019; Love & Crough, 2019). In this example, a list of topics was developed to spark curiosity in different aspects of protein science and cater to students in different programmes with different interests. Six topics were offered, and following voting, the two topics with the highest votes were taught. Students were required to write a reflection on their topic choice, and these reflections provided not only a direct insight into the motivation for choosing a particular topic but also their engagement in research to investigate topics prior to the democratic voting. The topic selection process resulted in students researching the available topics and selecting their choices based on their degree programme, future employment, or genuine interest. The student reflections from this approach were overwhelmingly positive with regards to being able to have an input to the curricula, for example:

> Loved being able to see a range of topics to choose from as it allowed me to grasp what topics exist overall in the topic of protein science. Choosing a topic meant an increased engagement and interest, and choosing questions for assessment meant I had to filter through what I know, didn't know, and what gaps I had in my knowledge.

In addition, students also valued being able to contribute to the assessment by design of multiple-choice questions which were used in practice quizzes and assessment for the course. This aspect of the partnership revealed the different learning strategies adopted/developed by students through the partnership activities, particularly in the design of multiple-choice questions and contribution to student assessment literacy (Love & Crough, 2019). Student comments suggested this partnership with students increased engagement: *"I enjoyed being able to choose different topics along with creating multiple-choice questions. I think it was a great way to keep us all engaged in our learning"*. The impact of this partnership with students to co-design aspects of the curriculum has increased student engagement, and this provides evidence for this teaching approach.

Partnerships between students and academics are becoming increasingly popular as academics recognise the importance of providing a voice for students, allowing them to have a say in their own learning (Mercer-Mapstone et al., 2017). Providing opportunities for students to make decisions about their own learning leads to an increased engagement, which, in turn, supports students in developing learning skills (Love & Crough, 2019). Having witnessed first-hand the engagement generated through student-staff partnerships, we are enthusiastic about establish a partnership with students in first-year biochemistry. Enhancing student motivation and engagement through a SaP teaching approach could complement the pre-COVID and COVID course improvements as previously mentioned. However, partnering with students in the first year raises some important questions. Do first year students have enough prior knowledge of the course content to contribute effectively to a staff-student partnership? How could we provide choices for students and ensure the key learning outcomes are met for successful transition into second- and third-year courses? What level of input would be required to inspire, motivate, and enhance student engagement in the course? A framework for exploring the level of student participation has been outlined by Bovill and Bulley (2011); however, student contributions in partnerships are both complex and contextual, and deciding on the appropriate level of student participation in course curricula needs to be based on both the context and the level of study (Bovill, 2013). This framework and consultation with current first year students will be used to determine the level of contributions for student partners. Understanding the student voice and what they perceive as important for their own learning will be an important aspect of establishing a staff-student partnership, particularly for our diverse student cohort, which contains a significant proportion of students from low socio-economic backgrounds and/or those that are "first in family" at university. The elements required to formulate a "genuine" partnership outlined by Matthews (2017) will also be considered in establishing the SaP in first-year biochemistry to ensure inclusiveness and negotiated power-sharing, leading to transformation through respect, reciprocity, and shared responsibility. Whilst we're not sure of the make-up of a partnership with first-year students, the tasks will involve active learning and reflection, which support lifelong and self-regulated learning.

CONCLUSIONS

In this chapter we have outlined the improvements made in a large first-year science course to improve student success. These improvements started prior to the pandemic, while others were developed for remote learning during COVID lockdowns. We have outlined further improvements to implement, post-COVID, as we move from traditional lectures to blended learning modes. The learning/teaching strategies outlined in the chapter are not applicable to our biochemistry course only, they are relevant to many first-year engineering courses which, like biochemistry, contain a large amount of content and complex threshold concepts. As we move out of the pandemic and students transition back onto campus, the face-to-face classes of blended teaching will need to focus on effective learning activities purposefully aligned with the curriculum content, and we believe that partnerships between students and staff will feature significantly in post-COVID teaching as students want a "voice" in the future of their learning.

ACKNOWLEDGEMENTS

We would like to acknowledge Christopher Allan and David Green (Learning and Teaching Consultants, Griffith Sciences Group, Griffith University) for their support with developing PebblePad workbooks. This project was supported by Griffith Sciences Blended Learning Funds. The use of student artefacts, reflections, and examples of student work has been approved by the Griffith University Ethics Committee under the following applications: 2017/1023 and 2021/649.

REFERENCES

Anderson, B. (2006). Writing power into online discussion. *Computers and Composition*, 23(1), 108–124.

Beins, A. (2016). Small talk and chit chat: Using informal communication to build a learning community online. *The Journal of Inclusive Scholarship and Pedagogy*, 26(2), 157–175.

Boekaerts, M., Pintrich, P.R., Zeidner, M. (2000). *Handbook of Self-Regulation*. San Diego: Academic Press.

Bond, J.B., Evans, L., & Ellis, A.K. (2011). Reflective assessment. *Principal Leadership*, 11(6), 32–34.

Bovill, C. (2013). Students and staff co-creating curricula: An example of good practice in higher education. In Dunne, L., & Owen, D. (eds.) *The Student Engagement Handbook: Practice in Higher Education*. Bingley: Emerald, pp. 461–475

Bovill, C., & Bulley, C.J. (2011). A model of active student participation in curriculum design: Exploring desirability and possibility. In: Rust, C. (ed.) *Improving Student Learning (ISL) 18: Global Theories and Local Practices: Institutional, Disciplinary and Cultural Variations*. Series: Improving Student Learning (18). Oxford: Oxford Brookes University: Oxford Centre for Staff and Learning Development, pp. 176–188.

Bovill, C., Cook-Sather, A., & Felten, P. (2011). Students as co-creators of teaching approaches, course design, and curricula: Implications for academic developers. *International Journal of Academics Development*, 71, 195–208.

Butler, D.L. (1998). A strategic content learning approach to promoting self-regulated learning by students with disabilities. In Schunk, D.H., & Zimmerman, B.J. (eds.) *Self-Regulated Learning: From Teaching to Self-Reflective Practice*. New York: The Guilford Press, pp. 160–183.

Chan, C.K.Y., & Lee, K.K.W. (2021). Reflection literacy: A multilevel perspective on the challenges of using reflections in higher education through a comprehensive literature review. *Educational Research Reviews*, 32, 100376, 1–18.

Cook-Sather, A., Bovill, C., & Felten, P. (2014). *Engaging Students as Partners in Learning and Teaching: A Guide for Faculty*. San Francisco, CA: Josey-Bass.

Crough, J., & Love, C.A. (2019). Improving student engagement and self-regulated learning through technology-enhanced student partnerships. In *Proceedings of the ICICTE*, July 4–6, Crete, Greece, pp. 112–124.

Dunlosky, J., Rawson, K.A., Marsh, E.J., Nathan, M.J., & Willington, D.T. (2013). Improving students' learning with effective learning techniques: Promising directions from cognitive and educational psychology. *Psychological Science in the Public Interest*, 14(1), 4–58.

Elkington, S. (2015). *Essential frameworks for enhancing student success: Transforming assessment in higher education – A guide to the Advance HE framework*. York, UK: Advance HE.

Eynon, B., & Gambino, L.M. (2017). *High-impact ePortfolio Practice: A Catalyst for Student, Faculty, and Institutional Learning*. Virginia, US: Stylus Publishing.

Freeman, S., Eddy, S.L., McDonough, M., Smith, M.K., Okoroafor, N., Jordt, H., & Wenderoth, M.P. (2014). Active learning increases student performance in science, engineering, and mathematics. *PNAS*, 111(23), 8410–8415.

Garrison, D.R., & Kanuka, H. (2004). Blended learning: Uncovering its transformative potential in higher education. *The Internet and Higher Education*, 7(2), 95–105.

Healey, M., Flint, A., & Harrington, K. (2014). *Engagement through Partnership: Students as Partners in Learning and Teaching in Higher Education*. York, Higher Education Academy. Retrieved November 21, 2021, from http://www.heacademy.ac.uk/resource/developing-undergraduate-research-and-inquiry

Hora, M.T., & Oleson, A.K. (2017) Examining study habits in undergraduate STEM courses from the situative perspective. *International Journal of STEM Education*, 4(1), 1–19.

Kift, S. (2009). *Articulating a transition pedagogy to scaffold and to enhance the first year student learning experience in Australian higher education: Final report for ALTC Senior Fellowship Program*. Strawberry Hills, NSW: Australian Learning and Teaching Council. Retrieved from http://transitionpedagogy.com/reports-and-resources/fellowship-report/

Kift, S. (2015). A decade of Transition Pedagogy: A quantum leap in conceptualising the first year experience. *HERDSA Review of Higher Education*, 2, 51–86. Retrieved from www.herdsa.org.au/herdsa-review-higher-education-vol-2/51-86

Kuh, G.D. (2008). *High-Impact Educational Practices: What They Are, Who Has Access to Them and Why They Matter*. Washington, DC: Association of American Colleges and Universities.

Larmar, S., & Lodge, J. (2014). Making sense of how I learn: Metacognitive capital and the first year university student. *The International Journal of the First Year in Higher Education*, 5(1), 93–105.

Love, C.A. (2020). Enhancing engagement by building relationships and a sense of community in an online biochemistry class. *Australian Biochemist*, 51(2), 13–14.

Love, C.A., & Crough, J. (2019). Beyond engagement: Learning from students as partners in curriculum and assessment. In *Proceedings of the 3rd EuroSoTL Conference*, Bilbao, Basque Country June 13–14, pp. 296–303.

Love, C.A., Crough, J., Green, D., & Allan, C. (2020). PebblePad workbooks for self-regulated learning: Study plans, learning strategies and reflection to promote success in first year. In Poot, A. (ed.) *Charting New Courses in Learning and Teaching: Case Studies from the PebblePad Community*, pp 42–48. US: Lulu.

Mann, K., Gordon, J., & MacLeod, A. (2009). Reflection and reflective practice in health professions education: A systematic review. *Advances in Health Sciences Education*, 14(4), 595–621.

Matthews, K.E. (2017). Five propositions for genuine students as partners practice. *International Journal for Students as Partners*, 1(2), 1–9.

McKenzie, K., & Schweitzer, R. (2001). Who succeeds at university? Factors predicting academic performance in first year Australian university students. *Higher Education Research and Development*, 20(1), 21–33.

Mercer-Mapstone, L., Dvorakova, L.S., Matthews, K., Abbot, S., Cheng, B., Felten, P., Knorr, K., Marquis, E., Shammas, R., & Swaim, K. (2017). A systematic literature review of students as partners in higher education. *International Journal for Students as Partners*, 1(1), 1–23.

Pantages, T.J., & Creedon, C.F. (1975). Studies of college attrition: 1950–1975. *Review of Educational Research*, 48, 49–101.

Pelletier, K., Brown, M., Brooks, C., McCormack, M., Reeves Arbino, N., with Bozkurt, A., Crawford, S., Czerniewicz, L., Gibson, R., Linder, K., Mason, J., & Mondelli, V. (2021). *EDUCAUSE Horizon Report, Teaching and Learning Edition*. Boulder, CO: EDUCAUSE.

Rashtchi, M., & Khoshnevisan, B. (2021). Developing online discussion boards to increase student engagement during the COVID-19 pandemic. *Dual Language Research and Practice Journal*, 4(1), 39–50.

Roberts, P. Maor, D., & Herrington, J. (2016). ePortfolio-based learning environments: Recommendations for effective scaffolding of reflective thinking in higher education. *Educational Technology & Society*, 19(4), 22–33.

Sebesta, A.J., & Speth, E.B. (2017). How should I study for the exam? Self-regulated learning strategies and achievement in introductory biology. *CBE – Life Sciences Education*, 16, 1–12.

Theobald, E.J., Hill, M.J., Tran, E., Agrawal, S., Arroyo, E.N., Behling, S., Chambwe, N., Cintron, D.L., Cooper, J.D., Dunster, G., Grummer, J.A., Hennessey, K., Hsiao, J., Iranon, N., Jones II, L., Jordt, H., Keller, M., Lacey, M.E., Littlefield, C.E., Lowe, A., Newman, S., Okolo, V., Olroyd, S., Peecock, B.R., Pickett, S.B., Slager, D.L., Caviedes-Solis, I.W., Stanchak, K.E., Sundaravardan, V., Valdebenito, C., Williams, C.R., Zinsli, K., & Freeman, S. (2020). Active learning narrows achievement gaps for underrepresented students in undergraduate science, technology, engineering, and math. *PNAS*, 117(12), 6476–6483.

Torrisi-Steele, G., & Drew, S. (2013). The literature landscape of blended learning in higher education: The need for better understanding of academic blended practice. *International Journal for Academic Development*, 18(4), 371–383.

Tsingos, C., Bosnic-Anticevich, S., & Smith, L. (2014). Reflective practice and its implications for pharmacy education. *American journal of pharmaceutical education*, 78(1), 1–10.

Zimmerman, B.J. (2000). Attaining self-regulation: A social cognitive perspective. In Boekaerts, M., Pintrich, P.R., & Zeidner, M. (eds.) *Handbook of Self-Regulation*. San Diego: Academic Press, pp. 13–39.

5 Replacing Laboratory Work with Online Activities

Does It Work?

Ivan Gratchev and Shanmuganathan Gunalan
Griffith University, QLD, Australia

CONTENTS

INTRODUCTION

Laboratories are commonly used in engineering and science to engage students in practical aspects of learning through hands-on activities and provide them with opportunities to test theoretical concepts from lectures and tutorials. Hodgson et al. (2014) noted that laboratories assist students with the development of practical, writing, and teamwork skills, which are important for their industry careers. During laboratories, students often work in small groups along with a lab instructor, creating a student-centred environment that promotes an appreciation of the lab and fosters metacognition (Matz et al., 2012). This interpersonal interaction can also lead to increased intrinsic motivation and enriched learning experiences, which are supported by face-to-face feedback in a timely manner (Hattie and Timperley, 2007). There are several advantages of in-person laboratories, such as giving students the ability to work with real equipment in authentic experimental settings and generate their own results (Brinson, 2015). In addition, laboratories tend to provide the most contact between students and the teaching team, which allows students to receive instant feedback on their performance.

Unfortunately, the COVID-19 pandemic led to the closure of universities following the public health authority advice to maintain social distancing. In response to

DOI: 10.1201/9781003263180-7

this, many engineering schools swiftly adopted an online teaching mode, where all teaching activities were delivered remotely. As a result, the online labs became the only option available for many engineering programmes during the pandemic; however, Qiang et al. (2020) suggests that online labs may become popular in the future, even after the COVID-19 restrictions have been fully lifted.

The literature shows that online labs can provide a cost-effective solution because students can access them from any location, thus eliminating the cost of lab space and material, and reducing the instructor's time (Brockman et al., 2020; Flint and Stewart, 2010). In such laboratories, pre-recorded videos can provide a step-by-step overview of real laboratory tests so that students can easily visualise, and become familiar with, the whole experimental process. Students can also replay these videos as many times as they like, an advantage that can save lab resources and avoid safety problems (Yesiloglu et al., 2021). The online labs can give students long-term access to the lab content, compared to the in-person two-hour-long labs (Brockman et al., 2020), which provides students with more time to process the knowledge and better prepare for lab assessments. The lab assessment may include online exercises and quizzes, which can be attempted multiple times as well. In addition, virtual labs can be part of online activities as they provide an environment in which experiments are conducted or controlled through computer simulation either locally or remotely via the Internet (Chan and Fok, 2009).

However, despite several advantages, online laboratories can discourage students from becoming familiar with physical instrumentation and real devices (Chan and Fok, 2009). In addition, the remote access denies valuable hands-on experience (Gamage et al., 2020), and it may discourage direct collaboration and interaction between students and the teaching staff. Other issues with online laboratories (Gamage et al., 2020) include the Internet connection, a lack of appropriate digital devices, and online cheating as students are not supervised. Recent experience indicates that students may not be satisfied with online labs when instructors use videos of experiments from video-sharing platforms, especially when it is not clearly related to the lab work (Yesiloglu et al., 2021).

This paper discusses the learning and teaching experiences obtained from two undergraduate courses, namely Engineering Mechanics (EM), and Soil Mechanics (SM), in 2020 and 2021. These courses are part of an undergraduate engineering degree programme offered at the School of Engineering and Built Environment (EBE), Griffith University, on two campuses (Gold Coast and Nathan campuses). Both courses teach students the fundamentals of EM, utilising a project-based approach in which students work on large projects throughout the whole trimester. The laboratories are an important component of each course because they help students better understand theoretical concepts, which can be rather difficult to visualise during lectures. The first part of this chapter discusses the structure of the EM course, the changes in the learning and teaching activities caused by the COVID-19 restrictions, student feedback, and academic reflection with a focus on student experience and engagement. The second part deals with the SM course in a similar way. As these two courses are offered to the student cohorts in their first and second years, the final conclusions provide an indication of the student experience with online laboratories at the early stage of their four-year undergraduate programme.

ENGINEERING MECHANICS COURSE

EM is a first-year course that develops a foundation and framework for many engineering disciplines such as civil, mechanical, and environmental. This course aims at developing a sound understanding of mechanics principles and problem-solving skills. EM contains a large amount of technical content, and consists of weekly lectures and workshops (Table 5.1). During the workshops, students perform a series of laboratory activities and discuss the obtained results with a laboratory instructor.

EM was redesigned in 2017 to better align with the experiential learning principles (Kolb, 2001, 2014). The course delivery was structured around a trimester-long project with scaffolding laboratory activities, which were performed during the workshops to facilitate a student-centred learning approach. This project-based approach built on real-world engineering problems provided students with authentic learning experiences (Crawley et al., 2014). The lab work helped students (1) to better understand the EM concepts, and (2) to connect theory with practice (Kober, 2015). The laboratory hands-on activities were also used to promote self-learning by observing and reflecting on different EM concepts (Table 5.1). Students employed a few physical models to experiment and learn about forces, moments, equilibriums, reaction forces, and deflections. Another series of experiments were designed to observe the truss member forces resulting from different nodal loads, the shear force, and the bending moment distribution in beams. During these lab activities, students

---

TABLE 5.1

Course Contact Hours before and after COVID-19 Restrictions and Workshop Activities

			Contact Hours	
	Hours per		**Before COVID-19**	
Activity	**Week**	**Total Hours**	**2017–2019**	**After COVID-19 2020**
Lecture	3	33 (11 weeks)	On Campus	Online (Live)
Workshop	2	22 (11 weeks)	On Campus	Pre-recorded videos and Online Drop-in-sessions

Workshop Activities	
Activity 1	Calculating the forces, moments, and components of forces using work board, plywood beam, string, weights, spring balance, pulley, and protractor.
Activity 2	Solving the equilibrium equations of a concurrent force system using work board, weights, strings, pulleys, and joint ring.
Activity 3	Understanding the free body diagram and calculating the support reactions of a non-concurrent force system using plywood beam, kitchen scales, and weights.
Activity 4	Performing truss analyses to find member forces and nature using cantilever and simply supported truss models.
Activity 5	Calculating second moment of area and deflection using PVC beams with different cross-sections, weights, dial gauge, and different support conditions.

Note that Griffith University uses a trimester system with 12 weeks of teaching in one trimester. One week is used as an employability week (no teaching).

---

worked in groups at their own pace, collected experimental data, and discussed it with the lab instructor, who provided students with instant feedback. The lab activities were followed by weekly quizzes that assessed student problem-solving skills and their ability to critically reflect on the laboratory results.

The student feedback on the face-to-face laboratories was mostly positive. It appears that the hands-on experience helped students better understand the EM principles and mathematical concepts. Students mentioned that they could (1) apply the fundamentals of physics and mathematics to analyse the equilibrium of simple systems under static loading, and (2) apply the concepts of sectional properties and internal force characteristics of beams to solve real-world engineering problems. Students were also able to predict, operate, and observe EM concepts, and reflect on the degree of agreement between estimation, calculation, and observation.

In 2020, due to COVID-19, the workshop activities were pre-recorded and uploaded to the course webpage so that students could watch the videos before each workshop. The workshop contact hours were utilised for drop-in sessions, where an instructor could answer student questions and clarify the content related to the pre-recorded videos. Many students watched the videos before each workshop and used the workshop time to interact with the instructor. However, there were some students who utilised the workshop contact hours to watch the videos.

RESULTS AND DISCUSSION

The implementation of the online lab activities seems to have engaged students in learning. This is evidenced by the data from the Student Evaluation of Course (SEC) survey (Table 5.2). The results for the Q1 question (*Overall, I am satisfied with the quality of this course*) indicate that in terms of the total score, the student experience with the online delivery (2020) was similar to the on-campus experience before COVID-19 in 2019. The Q2 question (*The workshop activities and tutorials in this course assisted my learning*) specifically targeted the workshop and laboratory activities. When the on-campus lab activities were replaced with the online mode, it was found that the score for student experience slightly improved for the student cohort at the Gold Coast campus (from 4 to 4.3), while the student experience score at the Nathan campus decreased from 4.4 to 4. It is still not clear why there was such a variation in the SEC score between two campuses; this can be related to the student demographics.

Student comments in the SEC survey give a better understanding on how the online delivery has influenced, motivated, and inspired students to learn. In 2019 (on-campus delivery), students enjoyed the course with the scaffolding lab activities. They appreciated the fact that these activities were hands-on, which helped them to understand some theoretical concepts. However, students also felt that the lab activities and quizzes had a great deal of content to cover within the two-hour workshops, and for this reason, the workshop activities were often rushed. As a result, they could not clearly understand some parts of the workshop material presented. According to the student feedback in 2020 (online delivery), students appreciated the online workshop videos and tutorial solutions. They felt that these videos helped them better engage in learning, and they even requested more videos to be produced.

TABLE 5.2

Student Feedback Regarding the EM Course and Laboratory Activities

Feedback	Campus	2019 (On Campus)		2020 (Online)	
		Score (out of 5)	Student Response Rate (%)	Score (out of 5)	Student Response Rate (%)
Q1	Gold Coast	4.5	28.1	4.3	28.4
	Nathan	4.7	37.7	4.7	28.1
Q2a or Q2b	Gold Coast	4.0	28.1	4.3	28.4
	Nathan	4.4	37.7	4.0	28.1

Note: Q1 *Overall, I am satisfied with the quality of this course.* Q2a (in 2019) *The workshop activities and tutorials in this course assisted my learning.* Q2b (in 2020) *The videos of workshop activities and tutorials in this course assisted my learning.*

The students commented that they enjoyed the workshop structure in which they could independently complete the activities at their own pace and attend the online workshop only when they needed help.

From a teacher's point of view, even though the online delivery of the lab activities provided students with a relatively good learning experience, there were still a few issues to be addressed. The main concern was that students were not able to experience the hands-on part of the laboratory. In addition, students could not effectively work as a group by using the online tools provided.

In 2021, the COVID-19 restrictions were relaxed, and the university allowed smaller classes to be held on campus. The workshops were delivered in a blended mode in which the online videos were available for students before online or on-campus drop-in-sessions. The students who could attend the on-campus classes (note that not all students could do so as some restrictions were still in place) were able to experience the laboratory activities in person. For such students, the pre-recorded videos served as additional learning tools.

SOIL MECHANICS COURSE

SM is designed to provide second-year civil and environmental engineering students with the fundamental knowledge of soil behaviour and opportunities to develop the critical skills necessary to apply this knowledge into practice. The weekly lectures deal with the theoretical aspects of SM while weekly tutorials scaffold student knowledge and assist students in developing problem-solving skills (Gratchev and Balasubramaniam, 2012). Five laboratories (scheduled fortnightly) are used to provide students with hands-on experience and connect theory with practice through practical work. Students are required to perform a series of laboratory tests, interpret and analyse the obtained results, and discuss their practical applications with a lab instructor (Gratchev and Jeng, 2018). Students commented that the labs were helpful in connecting the theory to real-world applications. They also noted that the face-to-face labs gave them some real exposure to the type of work done outside of university in this field.

In 2020, all teaching activities were shifted online. Online laboratory sessions were still held fortnightly while all lab experiments were pre-recorded and made available to students from the beginning of the trimester. Several videos created by the teaching team and technical staff explained a step-by-step procedure for each test, starting from the soil sample preparation using the relevant equipment, and to the analysis of the obtained results. The videos were edited to include text explanation of the key aspects of each experiment. This highlighted important details of the testing procedure, and helped direct learners' attention (Ibrahim et al., 2012). To maximise student attention, the videos were kept short, when possible, as according to Guo et al. (2014), the student engagement tends to drop when the video length becomes longer than six minutes. A total of 14 videos (Table 5.3) were recorded to cover the practical aspects of all lab sessions. The videos were uploaded to YouTube because this platform provided a variety of useful tools for both students and the teaching team.

The videos and online lab sessions were arranged in a way to replicate, to a certain extent, the experience that students would have during a face-to-face laboratory. Students were expected to watch these videos before each online lab session. The lab sessions were mostly used to discuss the results of each test and its practical application (Table 5.3), and they also served to answer students' questions. For each lab, a follow-up one-hour quiz, with a set of multiple-choice questions on the test procedure and short-answer questions related to data analysis, was given to the students on the next day. The quiz was available for 12 hours, and students could use all course

TABLE 5.3

The Content of Laboratory Sessions and Online Videos for Soil Mechanics in 2020

Tests		Online Laboratory Sessions
Lab 1	Dynamic cone penetrometer Vane shear test Pocket penetrometer test Water content test	Discussion of test results, data interpretation, and analysis. Discussion of practical application of all four tests.
Lab 2	Specific gravity Grain size (sieve) test	Focus on the use of test results to classify coarse-grained soils.
Lab 3	Proctor compaction test	Discussion of test procedure, data analysis, and application in construction.
Lab 4	Liquid limit test Plastic limit test Linear shrinkage test Constant head permeability test Falling head permeability test	Application of test results to classify plastic fine-grained soil based on soil plasticity. The use of soil permeability to estimate water flow under engineering structures.
Lab 5	Oedometer (consolidation) test Shear box test	Use of oedometer tests to estimate the time and settlement of soft soil under loads.

materials, including the video of each experiment. The previous experience with online quizzes suggested that this timeframe was sufficient for students to complete the test. To minimise online cheating, a pool of different problems was used to randomise quiz questions. Correct answers were provided on the following day to give students timely feedback. This also assisted students with their preparation for the course major assessment items such as Project 1 (25% worth) and Project 2 (35% worth).

RESULTS AND DISCUSSION

The data obtained from the YouTube statistics dashboard indicates a few viewing peaks around the day of the online lab session. It was clear that students watched the videos on the day before each session; however, the largest viewing peaks were observed on the day of the lab session, as well as on the day of the assessment quiz. This can be attributed to two facts: (1) students were required to watch the relevant videos prior to each online lab session; and (2) several questions in the quiz were related to the test procedure, which "forced" students to watch the videos before and during the quiz. No peaks were observed at any other time, suggesting that the students only watched the videos when absolutely necessary.

Although being unable to fully replace the student experience of face-to-face labs, the online lab sessions combined with the pre-recorded videos and online assessment quizzes provided students with the opportunities to engage in practical aspects of the course. Student feedback on the online videos was mostly positive, as they liked the opportunity to learn at their own pace and watch the videos at any time. However, there was on average about 10% of the student cohort who did not attempt the assessment quiz. The average mark of 7 (out of 10) for all five quizzes suggested that students acquired some knowledge of test procedures and developed skills to interpret the test results.

Student responses to the question regarding the level of satisfaction with the course are summarised in Table 5.4 for both campuses. It can be seen that, for Gold Coast campus the overall level of satisfaction (4.4/5) was the same for both years; that is, 2019 (on-campus) and 2020 (online). However, there was a noticeable change

TABLE 5.4
Student Satisfaction with the SM Course

Feedback	Campus	2019 (On Campus)		2020 (Online)	
		Score (out of 5)	Student Response Rate (%)	Score (out of 5)	Student Response Rate (%)
Q1	Gold Coast	4.4	27.7	4.4	33.3
	Nathan	4.8	41.7	4.0	9.7

Note: Q1 *Overall, I am satisfied with the quality of this course.*

in the score for Nathan campus, where it dropped from 4.8 to 4.0. It is noted that the relatively low student response rate, especially for Nathan campus, suggests that the overall score may not accurately represent the opinion of the whole student cohort. In addition, there might be other factors that influenced student satisfaction with the course.

In 2021, the lab sessions went back to a face-to-face delivery mode while the online videos were used as supplementary learning resources. As the videos were available before each lab, the time taken by the lab instructors to explain the test procedure significantly decreased, which allowed them to focus on the data analysis and practical aspects of each test. From an instructor's point of view, it is evident that the combination of online videos and face-to-face lab classes can enrich the student learning experience. Online videos of lab tests allow students to better understand the procedures and real-life applications while in-person labs can help them develop practical skills (Colthorpe and Ainscough, 2021). Therefore, for a blended mode, online lab videos can be effectively used as a preparation tool for other learning activities (Salter and Gardner, 2016).

The online labs, run in the same way as in 2020 during the COVID-19 restrictions, were also used in 2021 for offshore students and those students who could not attend the campus due to the lockdown or health concerns. This online lab arrangement proved to continue to be useful as it provided the students with an alternative way to engage in the practical aspects of the course.

CONCLUSIONS

COVID-19 has made academics explore and adopt different approaches towards teaching. They were required to replace in-person laboratory sessions with appropriate online activities. This was a challenging task as it required extra time, effort, and resources to produce new learning tools within a relatively short period of time. This paper presents and discusses the experience obtained for two engineering courses, in which face-to-face labs were replaced with pre-recorded videos and online activities. Based on the obtained results, the following conclusions can be drawn:

- Students understood the necessity of switching to online delivery and tried to adopt and engage in online activities. Students from different cohorts (first and second years of the four-year engineering undergraduate programme) provided similar feedback on the use of online activities, in particular online laboratories. Students generally liked the idea that online videos and other online resources were available from the beginning of the trimester so that they could access them at any time and work on them at their own pace.
- The follow-up quizzes used to assess the student knowledge and understanding of the laboratory work were designed in a way that not only "force" students to watch the videos, but also to help them understand the important aspects of each lab experiment. The videos prepared by the teaching team explained a step-by-step laboratory procedure so that students could

understand how the laboratory equipment was used and how the experimental results were obtained.

- The online sessions were used to discuss the details of each test and different approaches to interpret and analyse the lab data, as well as to answer student questions. However, compared to face-to-face labs, the major disadvantages of online labs included the lack of hands-on experience and the lack of communication between students, and between students and the teaching team.

- The online videos and other teaching resources developed during the pandemic can be successfully used as additional learning tools when face-to-face or blended delivery modes return. Watching online videos before a face-to-face laboratory can help students understand the experimental procedure and prepare them for the hands-on activities. It also allows the teaching team to save time during each lab, which can be used to focus on data analysis, discuss practical applications, and answer student questions.

As to the question put forward by the title of this chapter, the authors would like to note that online activities cannot completely replace the hands-on laboratory experience. However, pre-recorded videos of experiments combined with online sessions that facilitate student-teacher interaction and follow-up quizzes can provide a valid alternative to face-to-face labs, and can be effectively utilised when face-to-face sessions are not possible.

REFERENCES

Brinson, J. R. (2015). Learning outcome achievement in non-traditional (virtual and remote) versus traditional (hands-on) laboratories: A review of the empirical research. *Computers & Education*, 87, 218–237.

Brockman, R. M., Taylor, J. M., Segars, L. W., Selke, V., & Taylor, T. A. (2020). Student perceptions of online and in-person microbiology laboratory experiences in undergraduate medical education. *Medical Education Online*, 25(1), 1710324.

Chan, C., & Fok, W. (2009). Evaluating learning experiences in virtual laboratory training through student perceptions: A case study in Electrical and Electronic Engineering at the University of Hong Kong. *Engineering Education*, 4(2), 70–75.

Colthorpe, K., & Ainscough, L. (2021). Do-it-yourself physiology labs: Can hands-on laboratory classes be effectively replicated online? *Advances in Physiology Education*, 45(1), 95–102.

Crawley, E. F., Malmqvist, J., Östlund, S., Brodeur, D., & Edström, K. (2014). *Rethinking Engineering Education: The CDIO Approach*, Second Edition. New York: Springer.

Flint, S., & Stewart, T. (2010). Food microbiology – Design and testing of a virtual laboratory exercise. *Journal of Food Science Education*, 9(4), 84–89.

Gamage, K. A., Wijesuriya, D. I., Ekanayake, S. Y., Rennie, A. E., Lambert, C. G., & Gunawardhana, N. (2020). Online delivery of teaching and laboratory practices: Continuity of university programmes during COVID-19 pandemic. *Education Sciences*, 10(10), 291.

Gratchev, I., & Balasubramaniam, A. (2012). Developing assessment tasks to improve the performance of engineering students. In *23rd Annual Conference of the Australasian Association for Engineering Education 2012*.

Gratchev, I., & Jeng, D. S. (2018). Introducing a project-based assignment in a traditionally taught engineering course. *European Journal of Engineering Education*, 43(5), 788–799.

Guo, P. J., Kim, J., & Rubin, R. (2014). How video production affects student engagement: An empirical study of MOOC videos. In *Proceedings of the First ACM Conference on Learning@ Scale Conference* (pp. 41–50).

Hattie, J., & Timperley, H. (2007). The power of feedback. *Review of Educational Research*, 77(1), 81–112.

Hodgson, Y., Varsavsky, C., & Matthews, K. E. (2014). Assessment and teaching of science skills: Whole of programme perceptions of graduating students. *Assessment & Evaluation in Higher Education*, 39(5), 515–530.

Ibrahim, M., Antonenko, P. D., Greenwood, C. M., & Wheeler, D. (2012). Effects of segmenting, signalling, and weeding on learning from educational video. *Learning, Media and Technology*, 37(3), 220–235.

Kober, N. (2015). *Reaching Students: What Research Says about Effective Instruction in Undergraduate Science and Engineering*. Washington, DC: The National Academies Press.

Kolb, D. A. (2001). Experiential learning theory: Previous research and new directions. In *Perspectives on Thinking, Learning, and Cognitive Styles*, pp. 227–246. Routledge.

Kolb, D. A. (2014). *Experiential Learning: Experience as the Source of Learning and Development*. Pearson.

Matz, R. L., Rothman, E. D., Krajcik, J. S., & Banaszak Holl, M. M. (2012). Concurrent enrollment in lecture and laboratory enhances student performance and retention. *Journal of Research in Science Teaching*, 49(5), 659–682.

Qiang, Z., Obando, A. G., Chen, Y., & Ye, C. (2020). Revisiting distance learning resources for undergraduate research and lab activities during COVID-19 pandemic. *Journal of Chemical Education*, 97(9), 3446–3449.

Salter, S., & Gardner, C. (2016). Online or face-to-face microbiology laboratory sessions? First year higher education student perspectives and preferences. *Creative Education*, 7(14), 1869.

Yesiloglu, S. N., Gençer, S., Ekici, F., & Isik, B. (2021). Examining pre-service teachers' views about online chemistry laboratory learning experiences amid the COVID-19 pandemic. *Journal of Turkish Science Education*, 18, 108–124.

6 The Return to the Classroom after the Lockdown: New Challenges and New Education for a New Society

Jorge Membrillo-Hernández,
Patricia Caratozzolo, Vianney Lara-Prieto,
and Patricia Vázquez-Villegas
Institute for the Future of Education, Tecnológico de Monterrey, Mexico

CONTENTS

INTRODUCTION

In the course of human history, outbreaks of virulent diseases such as smallpox and the bubonic plague struck the massive, populated areas of the Roman Empire, the Chinese and Mayan civilisations, and half the continents of Europe and Asia. However, these outbreaks were contained within territorial limits (Human Population Through Time AMNH). From the middle of the 1700s, however, technology and medicine boomed with the advent of the Industrial Revolution.

The first transatlantic voyages were intended to transport goods. However, with the creation of flying machines, human beings were able to travel to other continents in a matter of hours, and despite World Wars I and II in the 1900s, the population began its great exponential growth (United Nations, 2019). During this period, globalisation

DOI: 10.1201/9781003263180-8

brought significant scientific advances, partly due to greater international collaboration in searching for solutions to global problems, such as global warming.

We recently experienced the worst pandemic in the last 100 years. Half of this world's population experienced a series of lockdowns that impacted migration and international tourism. More than 6 million lives were lost globally (Dong et al., 2020). An increase in poverty and food inequality, the accelerated deterioration of manufacturing, commerce, the economy, jobs, companies, the quality of medical care and schools, and the accentuation of digital and gender inequalities threatened the fulfilment of the United Nations' sustainable development goals (SDGs) put forward in 2015 (United Nations, 2015).

Despite this, the pandemic has also brought lessons. For example, fuel energy consumption and carbon emissions decreased in proportions not seen since 1900 (Le Quéré et al., 2021). The pandemic showed us that changes in behaviour could be made in an accelerated manner. With the return of normality, strategies must be adopted to continue these trends. Knowing and promoting the SDGs and their goals with a multidisciplinary vision will help us respond to these challenges.

Sustainable development is not assistance, charity, or volunteering, but a new development model based on literate citizenship with normative-ecological, altruistic, biospheric, eco-centric, utilitarian and future-relevant behaviour (Bliesner et al., 2014; Bybee, 2018). Growth projections for the next 100 years and the availability of resources required depend primarily on the decisions made now, which rely on the behaviour we adopt to address these problems.

This is possible if we invest in education for sustainability. That is, understanding how each act and each decision impacts the rest of the world. In a world that quickly changes, educational models must evolve simultaneously. Teachers are increasingly prepared and open to dealing with current problems. In this regard, Tencologico de Monterrey has implemented the Tec21 Educational Model based on four fundamental pillars: (a) challenge-based learning (CBL); (b) flexibility; (c) trained, inspiring teachers; and (d) memorable, integrated educational experiences.

This chapter describes experiences based on these four pillars. Although focused on experiences in science, technology, engineering, and mathematics (STEM) courses, the model can also be extended to other areas of study. The proposals explained here can guide a new design of classes and the generation of international collaborations in favour of a socially-oriented education (SOE).

SOCIALLY-ORIENTED EDUCATION (SOE)

Before the COVID-19 pandemic, there was already talk of this necessary change in human behaviour based on education. Neaman et al. (2018) introduced prosocial education, arguing that prosocial methods can contribute to behaviour in favour of the environment in environmental education. He captures it as an umbrella concept, involving many other ideas, including but not limited to those expressed in Figure 6.1.

In this sense, we are studying what type of approach works best for each context, and determining the psychological mechanisms behind the behaviour in favour of the environment due to a prosocial education (Neaman et al., 2018).

FIGURE 6.1   Concepts included in socially-oriented education.

After COVID, this SOE, defined as a process of change initiated at the end of the previous century and the beginning of this one, has greater scope and emphasis. Figure 6.2 illustrates, in the authors' opinion, the theories on which SOE is based, which, in turn, impact the decisions made in creating educational policies in contemporary education.

As can be seen, a generalised approach to the elements must be considered when designing sustainable teaching-learning strategies. Specific theories and areas of life that have supported the creation of the SOE are presented, and should be considered as a basis for the design and transfer of knowledge. In addition, the expected outputs are shown through digital technologies, such as the generation of skills and the resolution of contemporary challenges. However, to confirm that this change process continues to occur and accelerate, it is necessary to generate evidence on the progress of research in the area.

In response to this necessity for evidence, in 2021, the Socially-Oriented Interdisciplinary-SOI-STEAM research group was created at the Institute for Future Education of the Tecnológico de Monterrey; it comprises a collaboration between interdisciplinary experts in STEM education, social sciences, humanities, and the arts. One goal of the group is to promote sustainability as a transversal competence in study programmes, using education as both the objective and means. In this sense, the project is relevant for stakeholders, especially in the context of Latin America. It is based on the generation and transfer of knowledge, skills, attitudes, and behaviour, in a diverse, equitable, and inclusive way, for the achievement of the SDGs, especially in STEM (Membrillo-Hernández et al., 2021b).

Education's orientation towards society involves considering the complexity of social problems and proposing solutions based on theory, data, and technology

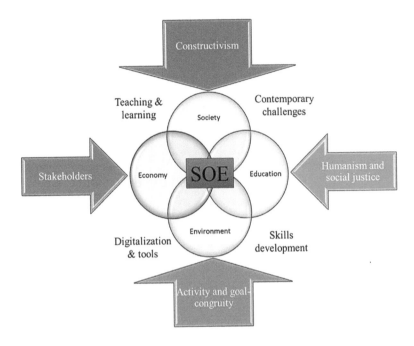

FIGURE 6.2 Scheme of what is included, and may be considered, in the development of empirical evidence that impacts modifications of the educational curriculum.

(Membrillo-Hernández et al., 2021b). For example, when searching for studies on "COVID" and "education" in SCOPUS, in the area of social sciences in the last two years, more than 18,000 empirical research documents appear, mostly (ca 80%) speaking of terms related to a pedagogical transition in favour of a change in education. It is mentioned that COVID was a catalyst for educational changes, mainly based on innovation and collaboration to solve challenges (Zhao, 2020). The main challenge was the sudden transition to remote work and study. However, much of the work is based on short-term solutionism. Therefore, continuity and planning are elements that have generally not been addressed (Brown and Finn, 2021) and are fundamental parts of the objectives of the research group. The following sections of the chapter present vital elements that can help STEM educators, from K12 to higher education, in strategic planning of classes and continuing their application through the continuous evaluation of the results obtained.

THE ACQUISITION OF DIGITAL AND OTHER SKILLS REQUIRED BY EMPLOYERS

During the lockdown, a lack of digital skills and a high workload were common elements of the transition to online classes. Anxiety and frustration were two primary emotions when students' professional futures remained uncertain, especially those with lower living standards, women, or students of applied sciences (Aristovnik et al., 2020). In addition, reports have been generated, for example, from the World Economic

Forum, which mentions the need to put holistic skills and educational pedagogies at the centre of learning. These skills have been categorised into global citizenship, innovation and creativity, technology, and interpersonal skills, within the framework of Education 4.0 in response to Industry 4.0. To accelerate this transition, they marked the trends that future schools must follow (Elhussein et al., 2020). These studies are critical for educational institutions to focus and apply changes in their policies for future similar events requiring emergency remote learning and virtual learning.

In this sense, the work that has been carried out includes meetings with other universities in Latin America to define action schemes in the implementation of innovative educational models within the framework of Education 4.0, using an experiential teaching-learning approach in the classroom and outside the school, as well as the interaction with companies in the training of students. This will enable students to develop a broader vision of the expectations of the industry, and to focus on the skills they must develop and the knowledge they require as a future workforce in the international market (Caratozzolo et al., 2021c).

Here, the theme of Generation Z that is now being formed contains a lot of relevance, since the educational structure of previous years has led to the appearance of barriers to the engagement of this generation, which is more connected and more careful regarding the environment and social issues, social and psycho-physical wellbeing (Songer and Breitkreuz, 2014; Adorno et al., 2021; Caratozzolo et al., 2021b). It has been found that, due to the accelerated technological progress of Industry 4.0, one of the primary skills these students must develop is digital literacy, which must be taught through the incorporation of state-of-the-art technological tools (Caratozzolo et al., 2021c). These skills can be developed through different pedagogical techniques of the active learning approach, which have also proven to be highly effective for developing sustainability thinking skills in engineering students (Caratozzolo et al., 2021d).

The pedagogical techniques of active learning arise with the objective that students develop transversal and disciplinary skills and competencies (González-Pérez and Ramírez-Montoya, 2022), such as negotiation, teamwork, and leadership, intercultural sensibility, adaptability, and a capacity to cope with uncertainty, ambiguity, and volatility in sustainable fields or technical skills in the area (Caratozzolo and Membrillo-Hernandez, 2021; Álvarez-Barreto et al., 2022). Challenge-based learning (CBL), problem-based learning (PrBL), and project-based learning (PBL) are some of these techniques. Figure 6.3 summarises the characteristics of these techniques; all focused on the student, which result in an improvement in academic learning (Membrillo-Hernández et al., 2019).

The results have indicated that, by using CBL, students acquire more knowledge. Students are required to have the ability to think critically and to solve problems, as outlined by the Accreditation Board for Engineering and Technology Inc. (ABET) criteria. They are exposed to real and open situations that demand a solution for which there is no pre-made response: for example, designing Net Zero Carbon 2050 strategies. In the process, through innovative technologies, they are exposed to situations of uncertainty, which, in turn, increases their resilience (Membrillo-Hernández et al., 2019).

In these collaborative projects, prior planning is necessary, and must define the skills that are sought to be highlighted, the didactic strategy that will serve to develop

Technique	Project based learning	Project based learning	Challenge based learning
Knowledge building process	Task-specific	Informative, self-directed	Working with teachers and experts
Project	Assigned	Designed	Real-world
Solution	Specific	Specific	Unspecific
Situation focus	Redefined problematic	Often fictional	Open
Product	Implementation of the solution	No real solution needed	Solution proposals
Assessment	Generated products	Knowledge application	Collaborative challenge tackling
Teacher's role	Facilitator and manager	Facilitator, guide, tutor, adviser	Designer, coach, co-researcher

FIGURE 6.3 Characteristics of active learning techniques. Adapted from Membrillo-Hernández et al. (2019).

the activity, transfer knowledge, the actions that students must do or complete to demonstrate learning and the evaluation rubrics that help to compare what was delivered by the students, as well as the means of feedback and the elements of continuous improvements, such as surveys. For this, the teachers require an adequate training programme (Membrillo-Hernández et al., 2021a).

In general, active learning strategies have been used in STEM courses to develop digital literacy skills, such as photo-visual ability, reproduction ability, branching ability, real-time critical thinking ability, and socio-emotional ability. It has been found that the strategy promoted a better understanding of scientific concepts in engineering subjects, a more remarkable ability to develop digital communication skills, and a greater awareness of creative thinking skills. In this sense, cognitive and technological tools generate better results for a taxonomy of knowledge that includes a cognitive process. However, accessibility, inclusion, and self-paced learning must be addressed (Caratozzolo et al., 2021a).

Collaborative Online International Learning (COIL) is another tool that has been used to promote the development of interpersonal skills and digital transformation of students. It enters the field of experiential learning as a cost-effective alternative with a high degree of effectiveness (Appiah-Kubi and Annan, 2020). Tecnológico de Monterrey has a Global Classroom, an international experience developed from the Vice-rectory for Internationalisation under the Global Shared Learning (GSL) initiative. In this initiative, a Tecnológico de Monterrey course is linked with one or more courses from an associated international university, through an international alliance, making use of technological tools to connect students, foster collaboration, and facilitate learning in intercultural environments.

Courses have been held in the Global Classroom with other universities in Latin America to potentiate students' learning experiences at both institutions by joining efforts to work on challenges in their respective careers. In addition to providing a memorable experience, the Global Classroom helps create international collaboration networks among the same students for future cooperation. These tools also help level the baseline to achieve equal job opportunities, considering that some students do not have the economic opportunity to travel to other countries as exchange students.

In this sense, the students who have had these experiences have given favourable opinions, since they have revealed a panorama of job opportunities and industries in other countries and widened the variety of problems they can help solve through their field of study (Caratozzolo et al., 2021c). Certain social paradigms can then be solved through education, for example, in making informed decisions when there is a need to migrate to other countries in search of employment or a better quality of life. These experiences allow them to increase their motivation and aspirations by opening their minds to a more global vision.

A similar programme is the implementation of flexible education; the use of "FIT" (Flexible, Interactive and Technologic) courses has also been a tool for the development of students' skills (Membrillo-Hernández et al., 2021a). These courses employ a hybrid, flexible format that allows both interaction with prominent professors in the same area of study at a national level, and the ability to work with colleagues on other campuses.

Another modality of learning sessions takes the form of "innovation weeks" (Tecnológico Week), where each learning experience is designed to take place over one week. They are intended to be immersion learning experiences, linked to the environment, in which students strengthen their skills (leadership, entrepreneurship, innovation, critical thinking, problem-solving, ethics, citizenship, global perspective, intellectual curiosity, passion for self-learning, collaborative work, communication, mastery of foreign languages and management of information and communication technologies) to become leaders who face the challenges of the current world (Tecnológico de Monterrey, 2021). In Innovation-Week (i-Week), students live real-life challenges through interactive and multidisciplinary activities with colleagues from different strategic sectors, disciplines, generations, campuses, and careers (Tecnológico de Monterrey, 2021). The Tecnológico Weeks allow companies to work with universities to solve their challenges. Studies have found that students can solve problems through these approaches by applying their knowledge in a fast and innovative way. It has been shown that students effectively adapt quickly to new technological tools (Lara-Prieto and Flores-Garza, 2022).

These weeks, offered each semester, present the opportunity to modify the learning sequences and improve the content according to the students' responses and opinions. This continuous improvement process makes it possible to detect areas of opportunity in the design of activities. At the same time, it allows teachers to maintain a constant learning paradigm to inspire students more and more. Results of these processes have shown a need to adjust assessment and advisory mechanisms (Smith et al., 2021).

This section has explained different ways in which students can acquire the necessary skills to succeed in their future employment. CBL can be implemented in online courses, face-to-face, or asynchronously. In the flexible format, there are tools or modalities that, using technology, allow students to have a memorable experience. In addition to this, the participation of stakeholders in the students' training and the design of the challenges will enable them to be more specific in the knowledge they must learn and the skills they must develop. In the following section, we make some recommendations on the design of new programmes and assessment systems to successfully develop skills, both in students and teachers.

CREATION OF NEW PROGRAMMES AND NEW EVALUATION SYSTEMS

Curriculum design based on STEM education has varied over the years. Considering stakeholders' interests for the generation of a workforce better adapted to the real world (Millar, 2020), the evaluation of the curriculum has been based on the measurement of the student's success. Although this success is subjective and context-dependent (Weatherton and Schussler, 2021), in the 1950s and 1960s, there was a quantitative measurement system based on scores of exams and knowledge tests, while from the 1970s to the 1990s, persistence, and factors such as motivation and goal pursuit were included in evaluation (Weatherton and Schussler, 2021). Yet, even with these theories, unrepresented students usually did not meet the established standards (Weatherton and Schussler, 2021). Those situations led to the urge for institutional change to ensure the success of all students (Knight, 2021). Between 1995 and 2008, disseminating curriculum and pedagogy, developing reflective teachers, enacting policy, and developing a shared vision were the most relevant topics of practice in STEM (Henderson et al., 2011). In recent years, STEM education has seen rapid changes, emphasising issues such as goals, policy, curriculum, evaluation, and assessment; evolving so quickly that some researchers suggest that a new revision will be necessary every five years (Li et al., 2020).

With the emergence of Industry 4.0 with its numerous benefits, it is necessary to enable the new workforce. The need for practical expertise and the implementation of digital technologies empowers future Industry 4.0 workers (Mian et al., 2020). In this regard, universities must adapt their educational programmes based on interdisciplinarity and skills development (Mian et al., 2020; Millar, 2020). Skills for sustainability, with its pillars, namely economic, social, and environmental (Sánchez-Carrillo et al., 2021), represent an opportunity to generate professionals who ensure enough production with fewer resources and prepare underprivileged students for a better job during this transition.

Interdisciplinarity, in its case, highlights the difficulty of measuring it. This interdisciplinarity must be intentional and specific, for which explicit strategies, feedback, and continuous improvement are required (Gao et al., 2020). On the one hand, being equitable (inclusive and fair) and diverse (origin and language), sustainable and digitised are today's main characteristics of STEM curricula. On the other hand, digital skills are the most promising skills that the workforce demands, as they will serve to put employees in charge of their own learning (Van Laar et al., 2017). In this sense, tools for quantitative mapping recognition of skills acquisition are necessary (Reid and Wilkes, 2016; Hill et al., 2020).

The new generations require continuous evaluation and recognition. Given digitisation and online classes, it is difficult to accurately picture the student's knowledge gain. The application of online exams is complex, since the information is on the internet and available to everyone. Therefore, the acquired knowledge must be challenged with problem-solving. It should be based not just on numbers but also on comparing multiple tools.

We have analysed several competency assessment systems using new educational technologies, finding that checklists and rubrics are more suitable, objective, and transparent in CBL classes (Membrillo-Hernández et al., 2021a). Another example is that the granting of badges to Global Classroom students generates motivation. To make a better quality system, blockchain technology as a quality assurance tool, in this aspect, could be helpful. The design and use of pre-tests and post-tests to measure learning gains are also of great importance. For this, the VALUE rubrics from the Association of American Colleges and Universities (AACandU) have been considered in courses with an active learning approach (Caratozzolo et al., 2021a). These rubrics can be modified according to the context of the field and the study environment. It is also essential to specify the form of evaluation from the beginning of the course and stick to it, since this reduces the student's anxiety levels and allows quality assurance in assessing the student's achievements. In this sense, specific values such as honesty, integrity, and fairness are also promoted (Whitelock et al., 2021). Educators oversee the design of the e-learning offer, with all the challenges it entails, from the recording of videos of the classes to limits in the access to online tools, which has resulted in a reconfiguration of their role. This is more relevant because faculty members are at a disadvantage compared to Generation Z students, who are digital natives, making educators digital migrants (Membrillo-Hernández et al., 2020).

Another evaluation system comprises the assessment centres in which internal and external teachers and other stakeholders (members of the community or companies) can evaluate students. In one of our studies, this modality was used to assess final degree students in engineering. In these centres, evaluators can ask students specific questions and receive feedback. This process can be carried out virtually, making it an equitable experience (Lara-Prieto and Niño-Juárez, 2021).

A change in the curriculum includes a difference in the mentality of the organisations, which must also maintain sustainable costs, effective financial planning, skilled staff, increased industrial partnerships, and advanced infrastructure, together with the inclusion of non-academic professional credentials that must necessitate a rethink of the curriculum (Keniry, 2020; Mian et al., 2020). However, the readiness of educators, management staff, and students was exacerbated by pandemics (Deák et al., 2021). This has given rise to the so-called "quantum leap" in the transition from using engagement in the curriculum to instilling success-oriented attitudes and behaviours (Larsen et al., 2021). To overcome those difficulties, third-party accreditation recommendations (QS, STARS); political, economic, social, technological, legal, and environmental (PESTLE) analysis, or strengths, weaknesses, opportunities, and threats (SWOT) analysis with analytic hierarchy process (AHP) addressing the viewpoints of stakeholders is recommended for universities to develop curricula and educational systems within a greater context (Mian et al., 2020; Deák et al., 2021).

CONCLUSIONS

In this chapter, tools are made available for the development of skills in the students of the new and future generations to solve contemporary challenges, with the

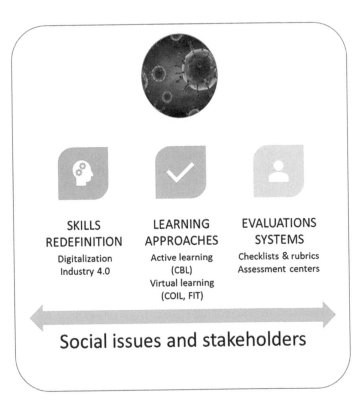

FIGURE 6.4 Strategies for new educational programmes towards a socially-oriented education (SOE).

participation of various stakeholders in the generation, transfer, and evaluation of knowledge. Quality evaluation strategies are proposed for digitised, personalised, and instant qualification. A summary of the recommendations we make in this chapter for educators towards a SOE is presented in Figure 6.4.

This period of pandemic showed that the breaking of the classroom is necessary. We must take the school outside the school. New educational approaches must respond to current challenges and positively transform society. SOE is crucial for sustainable development and vital for satisfying human needs (and minds), as are educational models that respond to particular and global realities. Global classrooms, training partners, and specialised mentoring are examples of strategies that mark the beginning of a new era in education.

REFERENCES

Adorno, D. P., Mallahnia, T., Koch, V., Zailskaitė-Jakštė, L., Ostreika, A., Urbaitytė, A., Punys, V., and Pizzolato, N. (2021). The BioS4You European project: An innovative way to effectively engage Z-generation students in STEM disciplines. *Education Sciences*, 11(12), 774. https://doi.org/10.3390/educsci11120774.

Álvarez-Barreto, J. F., Hernández, J. M., Chapa-Guillén, G. A., Larrea, F., and García-García, R. (2022). Internationalization in microbiology and bioengineering courses: Experiences between Mexico and Ecuador. In M. E. Auer, H. Hortsch, O. Michler, and T. Köhler (Eds.), *Mobility for Smart Cities and Regional Development—Challenges for Higher Education* (pp. 46–53). Springer International Publishing. https://doi.org/10.1007/978-3-030-93904-5_5

Appiah-Kubi, P., and Annan, E. (2020). A review of a collaborative online international learning. *International Journal of Engineering Pedagogy (IJEP)*, 10(1), 109. https://doi.org/10.3991/ijep.v10i1.11678

Aristovnik, A., Keržič, D., Ravšelj, D., Tomaževič, N., and Umek, L. (2020). Impacts of the COVID-19 pandemic on life of higher education students: A global perspective. *Sustainability*, 12(20), 8438. https://doi.org/10.3390/su12208438

Bliesner, A., Liedtke, C., Welfens, M., Baedeker, C., Hasselkuß, M., and Rohn, H. (2014). "Norm-oriented interpretation learning" and resource use: The concept of "open-didactic exploration" as a contribution to raising awareness of a responsible resource use. *Resources*, 3(1), 1–30. https://doi.org/10.3390/resources3010001

Brown, M. E. L., and Finn, G. M. (2021). Intra-COVID collaboration: Lessons for a post-COVID world. *Medical Education*, 55(1), 122–124. https://doi.org/10.1111/medu.14366

Bybee, R. W. (2018). *STEM Education Now More Than Ever*. Arlington, VA: National Science Teachers Association.

Caratozzolo, P., and Membrillo-Hernandez, J. (2021). Challenge based learning approach for Education 4.0 in engineering. In *SEFI 49th Annual Conference*. Berlin, Germany, 13–16 September 2021.

Caratozzolo, P., Alvarez-Delgado, A., and Sirkis, G. (2021a). Fostering digital literacy through Active Learning in engineering education. In *2021 IEEE Frontiers in Education Conference (FIE)*, 1–6. https://doi.org/10.1109/FIE49875.2021.9637304

Caratozzolo, P., Bravo, E., Garay-Rondero, C., and Membrillo-Hernandez, J. (2021b). Educational innovation: Focusing on enhancing the skills of Generation Z workforce in STEM. In *2021 World Engineering Education Forum/Global Engineering Deans Council (WEEF/GEDC)*, 488–495. https://doi.org/10.1109/WEEF/GEDC53299.2021.9657304

Caratozzolo, P., Friesel, A., Randewijk, P. J., and Navarro-Duran, D. (2021c, July 26). Virtual globalization: An experience for engineering students in the Education 4.0 framework. In *2021 ASEE Virtual Annual Conference Content Access*. https://peer.asee.org/virtual-globalization-an-experience-for-engineering-students-in-the-education-4-0-framework

Caratozzolo, P., Rosas-Melendez, S., and Ortiz-Alvarado, C. (2021d). Active learning approaches for sustainable energy engineering education. In *2021 IEEE Green Technologies Conference (GreenTech)*, 251–258. https://doi.org/10.1109/GreenTech48523.2021.00048

Deák, C., Kumar, B., Szabó, I., Nagy, G., and Szentesi, S. (2021). Evolution of new approaches in pedagogy and STEM with inquiry-based learning and post-pandemic scenarios. *Education Sciences*, 11(7), 319.

Dong, E., Du, H., and Gardner, L. (2020). An interactive web-based dashboard to track COVID-19 in real time. *The Lancet Infectious Diseases*, 20(5), 533–534. https://doi.org/10.1016/S1473-3099(20)30120-1

Elhussein, G., Leopold, A., and Zahidi, S. (2020). *Schools of the future: Defining new models of education for the fourth industrial revolution*. Future Skills Centre, Canada. Retrieved from https://fsc-ccf.ca/references/schools-of-the-future-defining-new-models-of-education-for-the-fourth-industrial-revolution

Gao, X., Li, P., Shen, J., and Sun, H. (2020). Reviewing assessment of student learning in inter-disciplinary STEM education. *International Journal of STEM Education*, 7(1), 1–14.

González-Pérez, L. I., and Ramírez-Montoya, M. S. (2022). Components of Education 4.0 in 21st century skills frameworks: Systematic review. *Sustainability*, 14(3), 1493. https://doi.org/10.3390/su14031493

Henderson, C., Beach, A., and Finkelstein, N. (2011). Facilitating change in undergraduate STEM instructional practices: An analytic review of the literature. *Journal of Research in Science Teaching*, 48(8), 952–984.

Hill, M. A., Overton, T., Kitson, R. R., Thompson, C. D., Brookes, R. H., Coppo, P., and Bayley, L. (2020). "They help us realise what we're actually gaining": The impact on under-graduates and teaching staff of displaying transferable skills badges. *Active Learning in Higher Education*. Sage journals. https://doi.org/10.1177/1469787419898023

Human Population Through Time | AMNH. (n.d.). American Museum of Natural History. Retrieved March 10, 2022, from https://www.amnh.org/explore/videos/humans/human-population-through-time

Keniry, L. J. (2020). Equitable pathways to 2100: Professional sustainability credentials. *Sustainability*, 12(6), 2328.

Knight, G. L. (2021). Delivering a positive outcome for STEM students – How TEF will that be?. *Higher Education Pedagogies*, 6(1), 37–40.

Lara-Prieto, V., and Flores-Garza, G. E. (2022). i-Week experience: The innovation challenges of digital transformation in industry. *International Journal on Interactive Design and Manufacturing (IJIDeM)*. https://doi.org/10.1007/s12008-021-00810-z

Lara-Prieto, V., and Niño-Juárez, E. (2021). Assessment center for senior engineering stu-dents: In-person and virtual approaches. *Computers and Electrical Engineering*, 93, 107273. https://doi.org/10.1016/j.compeleceng.2021.107273

Larsen, A., Cox, S., Bridge, C., Horvath, D., Emmerling, M., and Abrahams, C. (2021). Short, multi-modal, pre-commencement transition programs for a diverse STEM cohort. *Journal of University Teaching and Learning Practice*, 18(3), 05.

Le Quéré, C., Peters, G. P., Friedlingstein, P., Andrew, R. M., Canadell, J. G., Davis, S. J., Jackson, R. B., and Jones, M. W. (2021). Fossil CO_2 emissions in the post-COVID-19 era. *Nature Climate Change*, 11(3), 197–199. https://doi.org/10.1038/s41558-021-01001-0

Li, Y., Wang, K., Xiao, Y., and Froyd, J. E. (2020). Research and trends in STEM education: A systematic review of journal publications. *International Journal of STEM Education*, 7(1), 1–16.

Membrillo-Hernández, J., Ramírez-Cadena, M. J., Martínez-Acosta, M., Cruz-Gómez, E., Muñoz-Díaz, E., and Elizalde, H. (2019). Challenge based learning: The importance of world-leading companies as training partners. *International Journal on Interactive Design and Manufacturing (IJIDeM)*, 13(3), 1103–1113. https://doi.org/10.1007/s12008-019-00569-4

Membrillo-Hernández, J., García-García, R., and Lara-Prieto, V. (2020). From the classroom to home: Experiences on the sudden transformation of face-to-face bioengineering courses to a flexible digital model due to the 2020 health contingency. In *International Conference on Interactive Collaborative Learning*, 488–494. Tallinn, Estonia.

Membrillo-Hernández, J., de Jesús Ramírez-Cadena, M., Ramírez-Medrano, A., García-Castelán, R. M. G., and García-García, R. (2021a). Implementation of the challenge-based learning approach in Academic Engineering Programs. *International Journal on Interactive Design and Manufacturing (IJIDeM)*, 15(2), 287–298. https://doi.org/10.1007/s12008-021-00755-3

Membrillo-Hernández, J., Lara-Prieto, V., and Caratozzolo, P. (2021b). Sustainability: A pub-lic policy, a concept, or a competence? Efforts on the implementation of sustainability as a transversal competence throughout Higher Education Programs. *Sustainability*, 13(24), 13989. https://doi.org/10.3390/su132413989

Mian, S. H., Salah, B., Ameen, W., Moiduddin, K., and Alkhalefah, H. (2020). Adapting Universities for Sustainability Education in Industry 4.0: Channel of Challenges and Opportunities. *Sustainability*, 12(15), 6100.

Millar, V. 2020. Trends, issues and possibilities for an interdisciplinary STEM curriculum. *Science and Education* 29, 929–948. https://doi.org/10.1007/s11191-020-00144-4

Neaman, A., Otto, S., and Vinokur, E. (2018). Toward an integrated approach to environmental and prosocial education. *Sustainability*, 10(3), 583. https://doi.org/10.3390/su10030583

Reid, J., and Wilkes, J. (2016). Developing and applying quantitative skills maps for STEM curricula, with a focus on different modes of learning. *International Journal of Mathematical Education in Science and Technology*, 47(6), 837–.

Sánchez-Carrillo, J. C., Cadarso, M. A., and Tobarra, M. A. (2021). Embracing higher education leadership in sustainability: A systematic review. *Journal of Cleaner Production*, 298, 126675.

Smith, C., Onofre-Martínez, K., Contrino, M. F., and Membrillo-Hernández, J. (2021). Course design process in a technology-enhanced learning environment. *Computers and Electrical Engineering*, 93, 107263. https://doi.org/10.1016/j.compeleceng.2021.107263

Songer, A. D., and Breitkreuz, K. R. (2014). Interdisciplinary, collaborative international service learning: Developing engineering students as global citizens. *International Journal for Service Learning in Engineering, Humanitarian Engineering and Social Entrepreneurship*, 9(2), 157–170. https://doi.org/10.24908/ijsle.v9i2.5621

Tecnológico de Monterrey. (2021). Semana i. Accesado el 31 de enero de 2021. Disponible en https://semanai.tec.mx/es

United Nations. (2015). Transforming Our World: The 2030 Agenda for Sustainable Development. https://www.unfpa.org/resources/transforming-our-world-2030-agenda-sustainable-development

United Nations, Department of Economic and Social Affairs, Population Division. (2019). *World Population Prospects 2019: Highlights* (ST/ESA/SER.A/423).

Van Laar, E., Van Deursen, A. J., Van Dijk, J. A., and De Haan, J. (2017). The relation between 21st-century skills and digital skills: A systematic literature review. *Computers in Human Behavior*, 72, 577–588.

Weatherton, M., and Schussler, E. E. (2021). Success for all? A call to re-examine how student success is defined in higher education. *CBE—Life Sciences Education*, 20(1), es3.

Whitelock, D., Herodotou, C., Cross, S., and Scanlon, E. (2021). Open voices on COVID-19: Covid challenges and opportunities driving the research agenda. *Open Learning: The Journal of Open, Distance and e-Learning*, 36(3), 201–211. https://doi.org/10.1080/02680513.2021.1985445

Zhao, Y. (2020). COVID-19 as a catalyst for educational change. *PROSPECTS*, 49(1), 29–33. https://doi.org/10.1007/s11125-020-09477-y

7 Experiences of Fully-Remote Instruction for a Laboratory Course in Microwave Engineering

Berardi Sensale-Rodriguez
The University of Utah, Utah, USA

CONTENTS

INTRODUCTION

The COVID-19 pandemic has had a significant impact on our way of life. As an essential component of human society, education has not escaped these changes. Very likely, with respect to college-level education, one of the most significant changes that the pandemic has propelled has been a faster adoption of remote and online teaching methods. In this context, at the University of Utah, as pioneers in the flipped classroom method and in adopting technology for educational practices

DOI: 10.1201/9781003263180-9

(Furse and Ziegenfuss, 2020), the pandemic found us well poised for responding to these challenges. This chapter discusses our experiences in developing a fully-remote laboratory (lab) course on microwave engineering. This lab, although handled independently, is part of the Microwave Engineering course, an advanced technical elective in the broad thematic area of electromagnetics at the University of Utah. The lab course consists of five bi-weekly labs, spanning a 10-week period. These labs are intended to provide students with hands-on experience of microwave measurements employing a vector network analyser (VNA). As a second general objective, they aim to reinforce the concepts discussed in theory classes, particularly those of networks, network analysis, and S-parameters, as well as those behind microstrip lines, input impedance, and reflection coefficient. Furthermore, these labs are also intended to provide for an integrated design experience, ranging from a set of specifications and leading to a fabricated layout for important microwave components such as power dividers and filters.

In the context of an online class, these five labs were performed by the students remotely, asynchronously, and employing provided low-cost lab kits. These lab kits consisted of an open-source nano-VNA (https://nanovna.com/), four single-sided copper-clad FR4 boards, copper tape, 15 PCB edge mount SMA connectors, a 100 Ω SMD resistor (for Wilkinson power divider), an Exacto knife (for cutting the copper lines), and a soldering iron kit. Each student (20 students during the 2021 Fall semester) was supplied with a lab kit. Furthermore, students were provided with remote access to computers with Matlab and Advanced Design Software (ADS), an important CAD tool for the simulation and design of microwave circuits. The five labs in the Microwave Engineering course consisted of:

- Lab #1: *Introduction to the vector network analyser and one-port calibration,*
- Lab #2: *Design, fabrication, and test of microstrip lines,*
- Lab #3: *Design, fabrication, and test of a 2-way Wilkinson power divider,*
- Lab #4: *Design, fabrication, and test of a low-pass filter through the insertion loss method employing stubs,*
- Lab #5: *Design, fabrication, and test of a stepped-impedance low-pass filter.*

The development of these labs was based on our previous experience in setting up a remote lab for an introductory class in electromagnetics, which is described by Espinosa et al. (2021). In the following sections, we discuss the motivation and structure of each lab. We also discuss outcomes in terms of challenges and issues found by the students when working through the labs as well as student feedback in terms of their learning experience. From these observations, we envision strategies to further improve the student experience in future editions of the course.

DESCRIPTION OF THE LABS

LAB #1: INTRODUCTION TO THE VECTOR NETWORK ANALYSER AND ONE-PORT CALIBRATION

Motivation: the primary goal of this lab is for the students to gain a familiarity with basic vector network analysis. Furthermore, it is essential to acquire an understanding of the importance of calibration when performing measurements with a VNA. Calibration routines, which enable correction of systematic errors in the instrument, are implemented on firmware in most modern VNAs. However, it is crucial to gain an appreciation of how these techniques work in order to understand the VNA operation.

The usual procedure when making VNA measurements is to start by performing a calibration. This calibration procedure consists of measuring a series of known calibration standards. From these measurements, it is possible to construct a model of the microwave properties of the hardware internal to the VNA. This model is referred to as the "error model" or "error adapter". After generating this model, a routine in the VNA firmware mathematically removes the effects of the hardware's nonidealities from the measurement of the calibration standards. This enables correction for these imperfections in subsequent measurements. Errors that can be corrected through the calibration procedure must be systematic, thus intrinsic to the system and independent of the device under test (DUT).

For the case of a 1-port calibration, there are three system errors:

- **edf**: forward directivity error term resulting from signal leakage through the directional coupler on Port 1.
- **erf**: forward reflection tracking term resulting from the path difference between the test and reference paths.
- **esf**: forward source match term resulting from the VNA's test port impedance not being perfectly matched to the source impedance.

Shown in Figure 7.1a is a signal flow graph representing how these errors affect the measurement.

Standards typically employed in 1-port VNA measurements consist of an open $(Z = 0)$, a short $(Z = \infty)$, and a load (matched) termination $(Z = 50\Omega)$. From the independent measurements of each of the three standards, it is possible to extract the values of these three error terms (esf and erf). After these are computed, the S-parameters of any arbitrary DUT can be corrected. It is of note that additional calibration standards are needed for 2-port measurements. However, the ideas and concepts remain similar to those studied in this example. Therefore, this lab provides a useful introduction to the VNA, VNA measurements, and calibration process and, as such, is foundational to labs that come later in the course.

Description: in this lab, the students first collect all the necessary data for their calibration routine. For this purpose, they measure the three calibration standards with the nano-VNA. The VNA is initially uncalibrated, and the data are collected in their computers and exported as an s2p file. Once the data are collected, the students develop a calibration routine in Matlab; for this purpose, they are asked to derive the relations that the error coefficients in Figure 7.1a should follow in terms of the measured S-parameters from the calibration standards. Once the calibration routine is built, the students remeasure the three calibration standards and a nonterminated cable with the nano-VNA, and compare the obtained reflection coefficient (S_{11}) from their calibration routine with that from the internal calibration routine of the VNA.

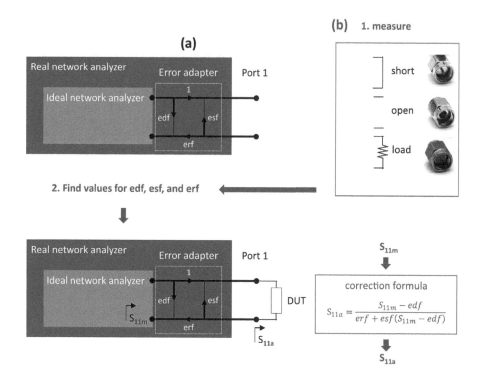

FIGURE 7.1   1-port VNA calibration process.

Theoretical and design aspects: critical to this lab is that students have previously (i) learned the definition and meaning of S-parameters; and that (ii) they have been introduced to signal flow graphs. As an activity to complete before the lab, students are asked to derive, using signal flowgraphs (rather than matrix formalisms), the input reflection coefficient for a 2-port network defined by a generic S-matrix ([S]) terminated by an arbitrary load (Z_L). The result of this question is that this input reflection coefficient (Γ_{in}) is given by:

$$\Gamma_{in} = S_{11} + \frac{S_{12}S_{21}\Gamma_l}{1 - S_{22}\Gamma_l}, \tag{7.1}$$

where Γ_l represents the reflection coefficient at the load and S_{ij} represents the S-parameters of the known network. From here, labelling the S-parameter measured by the VNA as S_{11m} and the S-parameter known for a calibration standard as S_{11a}, the following relation is derived between measured S-parameters and the known response of the standards:

$$S_{11a} = \frac{S_{11m} - edf}{erf + esf\left(S_{11m} - edf\right)}. \tag{7.2}$$

Three equations are derived here, one for each standard, with each of these equations linking the measured S-parameter with the known S-parameter of the standard ($S_{11a} = 0$ for the load, $S_{11a} = +1$ for the open, and $S_{11a} = -1$ for the short). From here, since we have three equations and three unknowns, the error correction terms (edf, erf, esf) can be obtained at each frequency of the measurement. Once these are known, the calibration routine just consists of employing Eq. (7.2) using the now known values of the correction terms as illustrated in Figure 7.1b.

LAB #2: DESIGN, FABRICATION, AND TEST OF MICROSTRIP LINES

Motivation: The purpose of this lab is threefold. The first aim of the lab is to reinforce the concept of microstrip lines as well as to provide an opportunity for the students to design, fabricate, and test the lines, thereby showing in practice that these lines behave as designed. A secondary goal of the lab is to provide an introduction/tutorial to ADS so that students become familiar with how to employ this CAD software for the simulation and design of more complex microwave components and circuits. Finally, the third objective of the lab is to lay the foundation for several upcoming lab sessions, which will deal with the design and construction of more complex microstrip-based microwave components. As such, this lab will serve as an introduction to microstrip transmission lines and the effect of nonidealities (such as fabrication errors, contacts, and discontinuities) on the lines' response.

Description: This lab starts with designing, that is, finding the width, for the top conductor in three microstrip lines of different characteristic impedance: $Z_0 = 20, 50,$ and $70\,\Omega$. These lines are then fabricated by pasting the copper film to the top of the provided FR4 board and cutting the copper film to the proper width. The student then solder SMA connectors to the edges of the board to connect to all three lines. Using the network analyser, measurement data, S_{11} and S_{21}, are then taken from these test microstrip lines. These data are vital, since they should be used in subsequent labs to properly model the properties of the substrate. In this lab, nominal parameters are assumed. Because of their intrinsic relation with the properties of a microstrip line (characteristic impedance and propagation constant), these material parameters of the substrate (e.g., permittivity, thickness, and loss tangent) and the parameters of the conductor (e.g., conductivity and width) can, in principle, be determined by properly fitting measured data to a model. This is performed in this lab through the "optimisation" tools in ADS. Furthermore, these optimisation tools are also employed to model the connectors and understand their effect in the measured S-parameters.

It is important to describe the fabrication procedure for the lines in more detail, since this is critical for subsequent labs and is a key aspect of the course. Unlike other remote microwave engineering courses wherein patterned structures are provided

(Iyer et al., 2021), an important aspect of this course is that the students perform all the fabrication and design at home. Illustrated in Figure 7.2 is the fabrication procedure, which is also detailed below:

i.   A section of copper tape is cut and pasted on top of the dielectric surface of the FR4 substrate.
ii. The layout of the three sections, with appropriate widths as per the design, is printed on vinyl or sticker paper at a 1:1 scale (see Figure 7.2a). This is then pasted on top of the copper tape pasted on top of the substrate.
iii. Using the Exacto knife, the sticker paper and copper film are carefully cut using a ruler as a guide (see Figure 7.2b).
iv. The excess regions are then carefully manually peeled off (see Figure 7.2c).
v.   Finally, the SMA connectors are soldered to the two ends of the three fabricated lines (middle pin) as well as to the ground plane in the bottom of the substrate (outer pins), as depicted in Figure 7.2d.

Theoretical and design aspects: A critical aspect of this lab and upcoming labs is to have an expectation of how results should look before doing the lab. This enables anticipating problems with either the design or fabrication of the microstrip microwave structures. For this purpose, students are asked at the beginning of the lab to design, assuming a 1.5 mm FR4 substrate (ε_r = 4.5), the three microstrip lines having characteristic impedances of (a) 20Ω, (b) 50Ω, and (c) 70Ω. For this purpose, it is recommended that the following design equations are employed (Pozar, 2011):

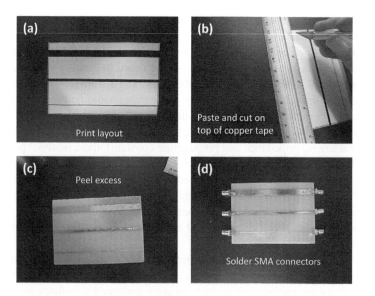

FIGURE 7.2   Microstrip fabrication process.

$$\frac{W}{d} = \begin{cases} \dfrac{8e^A}{e^{2A}-2} & for\ \dfrac{W}{d} < 2 \\ \dfrac{2}{\pi}\left[B-1-\ln(2B-1)+\dfrac{\varepsilon_r-1}{2\varepsilon_r}\left\{\ln(B-1)+0.39-\dfrac{0.61}{\varepsilon_r}\right\}\right] & for\ \dfrac{W}{d} > 2 \end{cases},$$

(7.3)

where:

$$A = \frac{Z_0}{60}\sqrt{\frac{\varepsilon_r+1}{2}} + \frac{\varepsilon_r-1}{\varepsilon_r+1}\left(0.23+\frac{0.11}{\varepsilon_r}\right).$$

$$B = \frac{377\pi}{2Z_0\sqrt{\varepsilon_r}}$$

(7.4)

Students are then asked to find an expression for the impedance at the input of the line (Z_{in}) analytically if these lines were terminated with a 50Ω load. This expression can be found from the general equation for input impedance as:

$$Z_{in} = Z_0^{line}\frac{Z_L + jZ_0^{line}\tan(\beta l)}{Z_0^{line} + jZ_L\tan(\beta l)}.$$

(7.5)

Then, assuming that these lines are connected to an instrument, that is the VNA, having a 50Ω port impedance (i.e., characteristic impedance), calculate the reflection coefficient $\Gamma = (Z_{in} - 50\Omega) / (Z_{in} + 50\Omega)$. Finally, sketch how this reflection coefficient will look as a function of frequency; i.e., what curve will Γ follow in the Smith chart space as βl is varied. When sketching this, it can be observed that the expected traces in the Smith chart space as varying frequency should be circles, as shown in Figure 7.3a. This should provide students with a physical understanding of the data, particularly S_{11}, as simulated/measured in this lab.

Furthermore, to include the effects of the SMA connectors, students are asked to consider a simple model for an SMA connector consisting of a series resistor and inductor, and a shunt capacitor. Initial values are provided for these elements; however, as previously discussed, these are later optimised to best fit the measured data. Depicted in Figure 7.3b is an illustration of the equivalent circuit for a line and connectors together with measured vs. simulated results before and after optimisation showing how these conductors, in particular, alter the measured response from theoretically expectations.

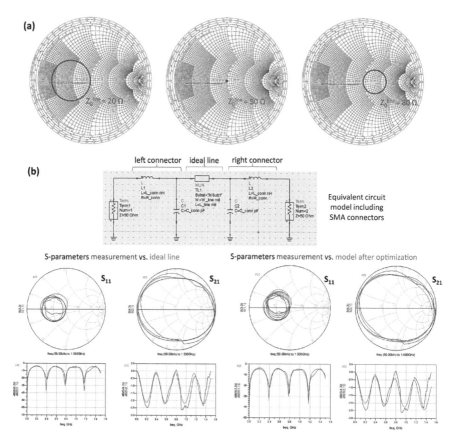

FIGURE 7.3 Response of different characteristic impedance transmission lines and equivalent circuit model containing SMA connectors.

LAB #3: DESIGN, FABRICATION, AND TEST OF A 2-WAY WILKINSON POWER DIVIDER

Motivation: The main goal of this lab is to employ the knowledge acquired during previous labs to construct a 2-way Wilkinson power divider. A 2-way Wilkinson power divider will split a signal into two with a nominal loss of 3-dB and isolation provided between ports. This geometry relies on the concept of a quarter-wave transformer. Furthermore, to match all three ports simultaneously, a resistor is connected between the second and third ports. This resistor basically isolates these two ports at the operating frequency of the splitter. The objectives of this lab are: (1) to design and layout a 2-way Wilkinson power divider using ADS; and (2) to build and test the divider. As a result of this lab, students also learn how to employ ADS to convert a schematic into a layout, which in this case is printed so as to physically realise the designed Wilkinson divider, as well as to use the "tune" feature in ADS to tune the microstrip lengths so that the Wilkinson operates as desired at a design frequency of 0.9 GHz.

Description: This lab starts by designing, analytically, a 2-way Wilkinson power divider at a design frequency of 0.9 GHz. After completing this initial design, a simple "ideal" circuit schematic is created for the circuit in ADS. The *S*-parameters are plotted and contrasted with theoretical expectations. The lab continues with creating a circuit schematic for a "real" Wilkinson divider; that is, by considering the bends of the waveguides (in all the arms) as well as the junctions between lines in the schematic layout as illustrated in Figure 7.4a. Once again, this is simulated, and associated S-parameters are plotted and contrasted with theoretical expectations. After simulation, students notice that the design is not precisely centred at the target design frequency. Therefore, in order to fine-tune the performance of the splitter at the design frequency, they are instructed to employ the "tune" feature in ADS. Once a good match is obtained, students proceed to create a physical layout of the microstrip circuit for the Wilkinson splitter. For this purpose, they use the "Layout" function in ADS. Both the size of the PCB and the size of the SMD resistor are taken into account when defining this layout. Shown in Figure 7.4b is a picture of the resulting physical layout, which is again simulated to validate results. Finally, the Wilkinson power divider is fabricated and tested following the same procedures as the previous lab. A picture of a fabricated PCB is shown in Figure 7.4c.

Theoretical and design aspects: before this particular lab, as well as for the next two labs, the students should already fully understand the theoretical knowledge regarding the operation and principles behind the operation of the component under design, in this case, a 2-way Wilkinson power divider. This is discussed in the earlier theory classes, and students are explicitly asked before commencing the lab to refer to the appropriate section of the course textbook (Pozar, 2011).

Lab #4: Design, Fabrication, and Test of a Low-Pass Filter through the Insertion Loss Method Employing Stubs

Motivation: Filters are central to microwave engineering and are employed in microwave circuits to select and control the transmission of signals of various frequency. These come in four primary groups named LP (Low Pass), HP (High Pass), BP (Band Pass), and BS (Band Stop); the names describe their functions. This lab aims to: (1) design and implement a low-pass filter in microstrip media employing ADS to simulate and optimise the result; and (2) build and test the filter. The prototype design for the filter is based on the "Insertion Loss Technique" (Pozar, 2011). In this lab, the students will learn how to scale the impedance and frequency of filter circuits and properly prototype filter designs for a given response. They will also learn how to work with stub elements to implement a filter.

Description: In this lab, the filter specifications are third order ($N = 3$) and maximally flat response. The filter should be designed for a cut-off frequency of 0.9 GHz. From here, an initial lumped-element filter prototype is determined, which then, by

(Continued)

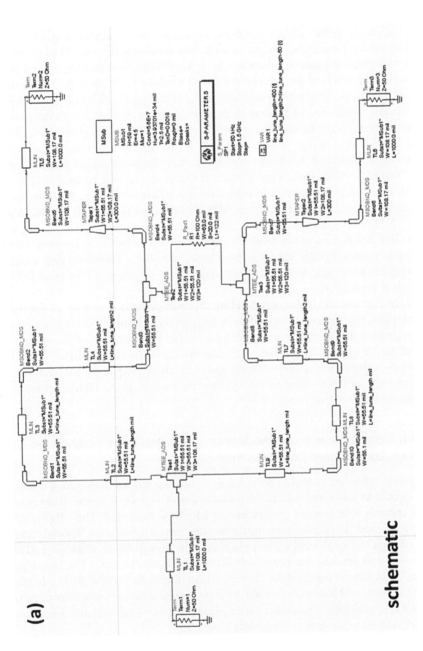

FIGURE 7.4 Wilkinson power divider.

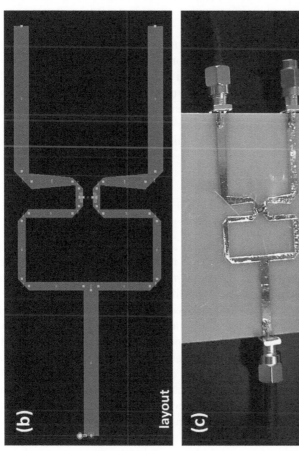

FIGURE 7.4 (Continued) Wilkinson power divider.

means of applying Richard's transformations and Koruda's identities, is implemented employing open-circuited stubs (Pozar, 2011). Students are asked to show and explain this design in detail, since this is an important aspect that links theory classes with this lab. Next, ADS is used to simulate the design of the filter and extract its physical layout. This is done by following the same procedures as in the previous lab. The design is adjusted to obtain the desired transmission over the passband and attenuation above the cut-off frequency. Once a physical layout is extracted, the filter is fabricated using the methods employed in the previous labs. Shown in Figure 7.5a is an example image of a fabricated filter. Finally, the frequency characteristics of the low-pass filter are measured. Illustrated in Figure 7.5b are the test results.

Theoretical and design aspects: Richard's transformations and Koruda's identities are central to this lab. Richard's transformations are employed to transform inductors and capacitors in the initial lumped-element LP filter prototype design into short-circuited and open-circuited stubs. Then, Koruda's identities enable physically separate stubs and transform series stubs into shunt stubs and vice-versa. Applying these concepts enables, therefore, to obtain a filter design consisting only of shunt open-circuited stubs, which is easy to layout with microstrips as pictured in the schematic in Figure 7.5c.

LAB #5: DESIGN, FABRICATION, AND TEST OF A STEPPED-IMPEDANCE LOW-PASS FILTER

Motivation: The goal of this lab is to design an LP filter employing the step impedance technique and to measure the frequency response of the filter. Central to this lab is also to understand the concept of stepped impedances and to utilise this concept to derive the microstrip implementation of the filter.

Description: In this lab, students are asked to design a stepped-impedance low-pass filter with the following specifications: cut-off frequency of 0.9GHz, third order, an impedance of 50Ω, and maximally flat response. For this design, it is assumed that the highest practical line impedance is 80Ω, and the lowest is 20Ω. Before undertaking the lab, students analytically perform the design. They then use ADS to simulate the design of the filter and extract its physical layout, fabricate a prototype for the filter using the methods employed in the previous labs and, finally, measure its frequency characteristics. Shown in Figure 7.6a–b is a sample image of the fabricated filter and exemplary results.

Theoretical and design aspects: Central to this lab is the concept of stepped impedances (Pozar, 2011), from which we can model a shunt capacitor as a small section of line of low characteristic impedance (Z_l) and a series inductor as a small section of line of large characteristic impedance (Z_h). This is illustrated in Figure 7.6c. As a result, from an initial filter design employing lumped elements, we can derive a microstrip design through substituting capacitors and inductors by transmission line sections of low and high characteristic impedance, respectively.

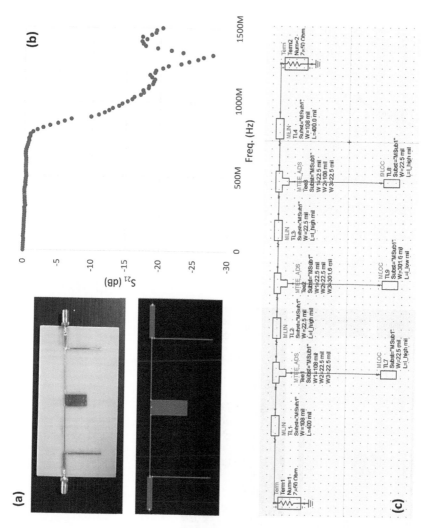

FIGURE 7.5 Low-pass filter implementation using stubs.

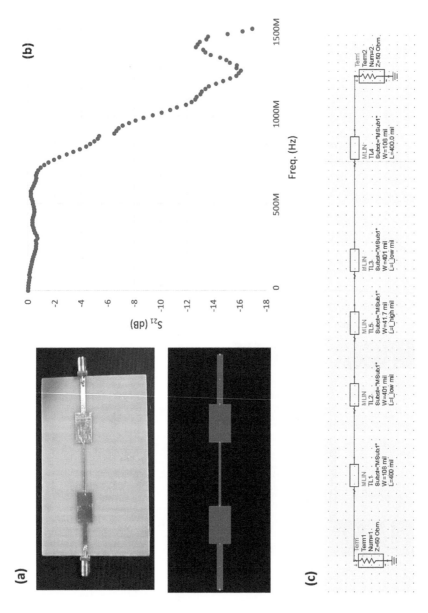

FIGURE 7.6   Stepped-impedance low-pass filter implementation.

OUTCOMES

Now that we have introduced the labs and explored the methods and concepts behind them, we discuss the outcomes of the course in terms of the challenges and issues found by the students when realising the labs and student feedback in terms of their learning experience. For this purpose, we conducted voluntary surveys from the students. Approximately 85% of the class (17 students) participated in these voluntary surveys. We also analysed the lab reports from all the students as well as communications between the students and teachers as a way to identify further challenges students faced when completing the labs.

GENERAL CHALLENGES AND OBSERVATIONS

When analysing the lab reports, we found a set of issues that repeated in several of those reports. For instance, in Lab #1, which required the development of a Matlab routine for VNA 1-port calibration, several reports contained issues resulting from coding errors or typos in the routines. These lead to inaccurate results. These issues were not identified even though results from the custom code for calibration and the intrinsic instrument calibration routine differed. An essential function of experimental lab education is to be able to identify and debug problems. In a traditional lab environment, the instructors constantly oversee the student's progress and aid them in making these types of observations/identifying mistakes.

In contrast, in a fully-remote environment such as for this course, it is the student's responsibility to identify these problems. This issue repeated in the second and third labs, but later on, it ceased taking place as students better understood the lab dynamics. Another two general observations that we observed across the lab reports are: issues with following instructions, particularly pre-lab instructions, and time management. We believe that the asynchronous, remote nature of the course significantly impacted this. As students have a two-week window for working the labs and writing their reports, it was suggested that they start as soon as possible. However, in general, we observed that students would often reach out with questions just a couple of days before, or only on, the report's due date. In contrast to a synchronous course, an asynchronous course requires students to decide how to manage their own time. We observed that this was a challenge for many students.

Furthermore, when analysing the lab reports, we observed in several cases that foundational pre-lab activities and questions were sometimes not addressed in the reports, even though these were explicitly listed as things to do as part of the lab guidelines. In general, reports from students not following the lab guidelines, and in particular, not completing these pre-lab activities, tended to present more issues with regard to incorrect results. Simply put, if a student does not understand the expected result of an experiment, they will not recognise an incorrect result if an error occurs. Once again, in a regular lab setting, the instructor is on the spot to identify issues, explain concepts, and explain whether there are problems with results or methods and why. In a remote lab setting, the student often needs to take the leading role and initiate contact with the instructors. It is of note that the instructors had regular office hours (virtual and in-person) and were available to hold in-person lab sessions based on students' requests.

RESULTS FROM LAB FEEDBACK

The first question on the voluntary lab feedback questionnaires was: "How much time did it take you to perform the lab?" The motivation behind this question is to understand not just how much time was spent, but also to guide the lab structure for future editions of the course. In a traditional lab setting, the expected total dedication per lab (including report writing) is expected to be around six hours. Shown in Figure 7.7 are the results from the survey showing the average and standard deviations of this time dedication by lab, as well as the breakdown data for each of the labs. We observe that in the first, fourth, and fifth labs, the average time spent per student approaches this 6-hr level (six hours over two weeks).

Furthermore, the standard deviation is relatively small. However, in the second and third labs, the average time spent was substantially more significant, and it varied greatly within the class, with some students declaring spending as much as 20–30 hours in these labs. It is noteworthy that these two labs are the ones introducing CAD software and simulations as well as dealing for the first time with fabrication and measurement of microstrips. This brings us to the answers to the next question, "What were the greatest challenges/issues that you encountered?" Herein, students found it, in general, challenging to learn how to get started with ADS, and also encountered difficulty in fabricating microstrip lines and soldering connectors. This is very important feedback that shows that structuring the lab to last only five sessions might be too short a time. For the course's subsequent editions, we see value in having a dedicated lab only on CAD simulations, serving as a tutorial, and splitting the experimental portion into an independent lab.

Regarding the question, "Based on your experience, what do you wish you have known before doing the lab?", two common themes observed in the answers were: to have understood how the component under analysis worked before doing the lab, and to have known what results to expect. As mentioned, the lab was not independent of the theory classes in this microwave engineering course; however, the course was remote and asynchronous. Therefore, it was, in part, up to the students to choose the pace at which to work. This seems to be an overall challenge in remote and online education that can be magnified in the context of remote laboratory instruction. However, by asking students to work and submit pre-lab assignments before having access to the actual lab, it might be possible to ensure that students understand the theoretical concepts and are clear on expected results before conducting the experiments of the lab. Providing a more "rigid" course structure, breaking up assignments, and having more set deadlines can be beneficial from this perspective.

Finally, with respect to the last two questions, "What are the most important/interesting things that you think you learned by doing this lab?" and "How do you feel that this lab relates to the course?", a common theme was that students found value in applying the concepts discussed in the theory class in a physical context. For example, a particular comment of one student was that the labs provided an idea of what a microstrip actually is. By creating, touching, and measuring, abstract concepts and elements do become real; this is a vital role of laboratory education, even in a remote setting. A second common theme was the value students found in the way the labs reinforce theoretical concepts; for instance, how impedance and width on a

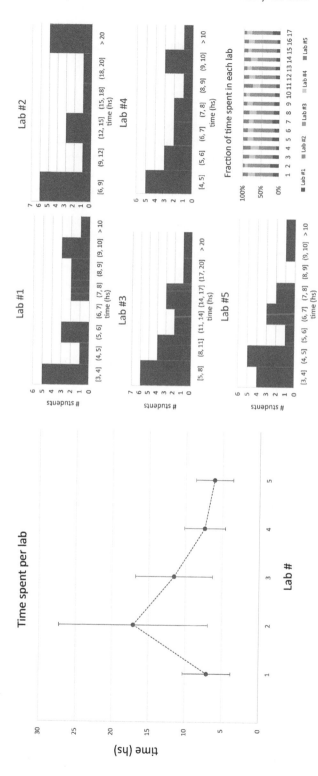

FIGURE 7.7 Declared time spent per lab during the course.

microstrip are related, or the design flow for a low-pass filter. Overall, students observed that they had a better understanding of the course topics after performing the labs.

CONCLUSIONS

Although challenging in many aspects, providing an effective, fully-remote laboratory course on microwave engineering where students can undergo the complete design-fabrication-test process of simple microwave components is possible thanks to current low-cost instrumentation. In this chapter, we discussed an example of such a course, consisting of five lab practices performed remotely, asynchronously, employing low-cost lab kits provided to each student. Overall, the feedback from the students was very positive. However, common intrinsic challenges of online and remote education were observed, which, although not general to the class, show that improvement is needed so that the learning experience is effective for all students. Splitting some of the labs into shorter exercises, focusing labs on only one topic, and conveying the importance of pre-lab questions and exercises could, in part, address some of these problems.

REFERENCES

Espinosa, H. G., Khankhoje, U., Furse, C., Sevgi, L., & Rodriguez, B. S. (2021). Learning and teaching in a time of pandemic. In K. T. Selvan & K. F. Warnick (Eds.), *Teaching Electromagnetics: Innovative Approaches and Pedagological Strategies* (pp. 219–237). CRC Press.

Furse, C. M., & Ziegenfuss, D. H. (2020). A Busy Professor's Guide to Sanely Flipping Your Classroom: Bringing Active Learning to Our Teaching Practice. *IEEE Antennas and Propagation Magazine*, 62(2), 31–42.

Iyer, A. K., Smyth, B. P., Semple, M., & Barker, C. (2021). Going Remote: Teaching Microwave Engineering in the Age of the Global Pandemic and Beyond. *IEEE Microwave Magazine*, 22(11), 64–77.

Pozar, D. M. (2011). *Microwave Engineering*. John Wiley & Sons.

8 Geo-Engineering Teaching and Learning during the COVID-19 Lockdown

The University of Auckland Experience

Martin S. Brook, Rolando P. Orense, and Nick P. Richards
University of Auckland, Auckland, New Zealand

CONTENTS

DOI: 10.1201/9781003263180-10

INTRODUCTION

The spread of the novel coronavirus COVID-19 has impacted every economic and social sector across the globe (Carruthers, 2020). At both undergraduate and postgraduate levels, higher education institutions in New Zealand, like elsewhere, have opted to cancel all face-to-face classes, including laboratory and other learning experiences, and have directed that those classes must be held online to help prevent the spread of the virus. With the shift to Emergency Remote Teaching, academic staff had to rapidly modify their teaching materials so that students can study online. On-campus teaching at a range of institutions, out of necessity, has had to be modified. Academics have had to develop innovations and facilitate flexible ways to offer both practical and theoretical components of courses. This has meant alterations to both formative and summative assessments, as well as content. Academics with teaching responsibilities have had to upskill and quickly familiarise themselves with online learning platforms, including additional, unfamiliar administration processes. The move to online teaching has forced some academics to re-evaluate their curricula against a backdrop of rapid and ubiquitous technological changes (Roy and Metson, 2021).

In New Zealand, the Government closed the country's borders to non-citizens and non-residents on 19 March 2020, just after the start of the academic year (Figure 8.1), and for the next two years, has followed an elimination strategy, with some of the strictest lockdowns globally (Eleanor, 2020). On 23 March 2020, the Government introduced a four-tier alert level system, which gave the country 48 hours' notice that the citizens would be locked down, aside from access to essential services such as pharmacies and supermarkets. At the University of Auckland, Semester 1 started three weeks earlier on 2 March 2020, so this lockdown had an immediate effect. Over the course of the next two academic years, the country fluctuated between alert levels that created different restrictions. COVID-19 cases that have broken through the international border have generally emerged in Auckland, the country's business and economic centre. Thus, Auckland has been more affected by lockdowns than have the other universities in various regions of New Zealand (Figure 8.1).

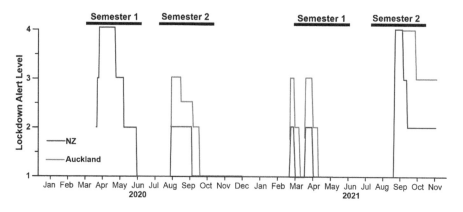

FIGURE 8.1   Semester dates at the University of Auckland and lockdown levels.

The first three weeks of Semester 1 2020 had just been completed when New Zealand went into full lockdown in March 2020. As a result, the remaining weeks were shifted to online mode. In mid-2020, the country opened again, so Semester 2 2020 had the first three weeks on-campus before moving to online teaching from Week 4 until the end of the semester. In 2021, Auckland went into lockdown in February, affecting the first two weeks of lectures, but the remaining weeks of the semester were in face-to-face mode. All final examination assessments in 2020–2021 were undertaken online.

This chapter outlines how the engineering teaching staff in two academic units within the University of Auckland have adapted to remote, "emergency" teaching. We report how the Department of Civil & Environmental Engineering staff adapted their teaching for two compulsory undergraduate geotechnical engineering papers, CIVIL 221 (Geomechanics 1) and CIVIL 322 (Geomechanics 2), during the lockdown. We also report how the School of Environment's Engineering Geology teaching staff adapted to teaching an undergraduate and postgraduate paper, EARTHSCI 372 (Engineering Geology) and EARTHSCI771 (Advanced Engineering Geology), respectively. At the University of Auckland, each semester runs for a total of 14 weeks, including a 2-week break that splits each semester into two halves.

DEPARTMENT OF CIVIL & ENVIRONMENTAL ENGINEERING TEACHING

DEGREE PROGRAMME

The Degree of Bachelor of Engineering (Honours) [or BE (Hons)], majoring in Civil Engineering, is a 4-year undergraduate degree taught in the Department of Civil and Environmental Engineering. After an internal review and consultation with the industry and various stakeholders, the degree programme underwent a major restructure. The new BE (Hons) programme was implemented in 2020 for incoming (Part I) students. This section follows the adjustments made in teaching two compulsory geomechanics courses in the "old" curriculum to the student cohort affected in 2020–2021 when Auckland went into lockdown and teaching shifted from face-to-face setting to online teaching.

GEOMECHANICS COURSES: USUAL DELIVERY

In the "old" BE (Hons) programme, three compulsory 10-point (pt) courses (CIVIL 220: Introductory Engineering Geology; CIVIL 221: Geomechanics 1; and CIVIL 322: Geomechanics 2) provide the core geotechnical knowledge thought to be essential for all graduates in Civil Engineering or Environmental Engineering. They are also intended as a preliminary to the geotechnical engineering elective courses that are offered in Parts III and IV.

CIVIL 221 (Geomechanics 1) focuses on the basic concepts and principles governing the mechanical behaviour of soil, including phase relationships, permeability and seepage, effective stress principle, soil strength, compressibility, and basic stability analysis. In Semester 2, 2020, 307 Part II, students were enrolled in the course.

However, the continuation CIVIL 322 (Geomechanics 2) course covers stability analysis in geotechnical engineering, slope stability, soil pressures on retaining structures, bearing capacity, and consolidation and settlement. There were 266 and 300 students enrolled in this Semester 1 course in 2020 and 2021, respectively.

Being both 10 pt papers, each of the two courses involve, on a weekly basis, 3 hours of lectures and tutorials (approximate ratio of roughly 2:1), 1 hour of reading and thinking about the content, and 3 hours of work on assignments and/or test preparation.

CIVIL 221 has three 2-hr laboratory (lab) sessions (Lab 1: soil classification + particle-size distribution; Lab 2: flownet construction in seepage tank; and Lab 3: unconsolidated undrained triaxial test). On the other hand, CIVIL 322 has two 2-hr lab sessions (Lab 1: consolidation test and soil compressibility; and Lab 2: triaxial compression test on dry sand).

In 2020–2021, the final examination in CIVIL 221 was worth 50% (other assessments included lab reports, individual/group projects, and tests), whereas the final examination was worth 55% in CIVIL 322 (other components included lab reports, group projects, quizzes, and tests).

GEOMECHANICS COURSES: ADAPTATION TO ONLINE TEACHING

All final assessments in 2020–2021 were online, with assessments in Semester 1 2020 undertaken over a 24-hr period. In the other three semesters, assessments ran for 2-hr periods (+ additional 30 mins as reading time and for uploading answer sheets) for these two 10 pt courses. Tests of 50 mins duration were also conducted online.

Lectures

With the shift to online teaching, the civil engineering lecturers uploaded their lecture materials to Canvas, the university's Learning Management System (LMS). They supplemented these materials with lecture recordings, mostly new recordings split into smaller "chunks" of 20–30 mins duration. In a few cases, the previous year's lecture recordings were also used. Short video clips, mostly taken from YouTube, supplemented some lectures to emphasise important concepts. The lecture materials and recordings were released to students in stages following the predetermined course outline. Also, students could access the appropriate course materials only at designated times (generally based on the scheduled lecture hours, per the students' course timetable); this encouraged the students to follow their lecture schedules and simulate the campus environment.

Tutorials were generally conducted via Zoom, in most cases, as part of the lecture recordings. When done separately, example problems were prepared in advance, and solutions were discussed. These sessions were split into smaller chunks of 15–20 mins duration. However, because of the large size of the class (roughly 300+ students), these Zoom sessions were recorded for the benefit of those who could not attend. Online catch-up sessions and regular office consultations were also conducted regularly via Zoom.

The lecturers made full use of Piazza, an intuitive platform linked to Canvas, to efficiently manage class question and answer Q&A sessions. The platform acted as a moderated online discussion forum, where students could post questions and collaborate to edit responses to these questions; at the same time, lecturers could also answer questions, endorse student answers, and edit or delete any posted content. Lecturers generally attempted to answer students' questions within 24 hrs from the time of student posting, and even faster as the need arose.

To facilitate retrieval of online learning materials and resources, these were stored in a structured, easy-to-access way via Canvas Modules. Using this tool, the lecturers could structure the course contents either on a daily or weekly basis or through similar units or topics. This approach simplified how students navigate through the courses while ensuring a sequential flow of content.

Laboratory Sessions

The biggest challenge with the shift to online teaching was the conduct of the lab sessions. These lab sessions were designed for the students to "feel" the soil and to familiarise themselves with experiment aims and procedures. Under normal circumstances, these sessions were conducted in small groups (around 25–30 students per session) and held at the Faculty of Engineering's Multi-Disciplinary Learning Space (MDLS).

With the shift to online teaching, the lecturers had to improvise strategies to illustrate the required concepts. One method adopted was an explanation of the conduct of the tests in detail as part of the lecture materials (either as recordings or as short memos). For example, in lab sessions involving triaxial testing, the explanation of the test procedure was supplemented by videos, mostly from YouTube. Students were then provided with example "raw" test data to explain how the relevant parameters from the tests were obtained.

At the end of each lab session, the students were provided with "experimental results" and were required to make the appropriate analyses. The students then submitted their lab journals, which should reflect how they processed the data and how they estimated the parameters needed through graphical and/or analytical approaches.

SCHOOL OF ENVIRONMENT ENGINEERING GEOLOGY TEACHING

Degree Programmes

Engineering Geology teaching occurs at both undergraduate and postgraduate levels. The Degree of Bachelor of Science, majoring in Earth Science, is a 3-year undergraduate degree, worth 360 points (i.e., 24 × 15 pt papers; 8 papers/year), and is taught in the School of Environment. The Earth Science major was restructured in 2019 around a "spine" of compulsory courses that contain the typical elements required by a graduate geologist working in the New Zealand industry. While there are two compulsory papers in the first year, there is only one compulsory course in the second and third years. The rest of the BSc Earth Science major is composed of a suite of elective papers, including typical geological themes; e.g., EARTHSCI 372 Engineering Geology. While the BSc Earth Science programme includes a broad range of geology-related papers typical of a New Zealand BSc Earth Science major, the 15-pt EARTHSCI 372 Engineering Geology is the one paper within the BSc that has a work-integrated learning (WIL) ethos. Recognised within higher education, WIL refers to the range of education and training opportunities that together aim to improve the career prospects of graduates and their transition to work by providing valuable practical experiences informed by university study (Patrick et al., 2009). In addition, the Master of Engineering Geology is a 180-pt, 18-month degree. It comprises a 90-pt thesis and 6 × 15 pt taught papers, of which EARTHSCI 770 (Engineering Geological Mapping), EARTHSCI 771 (Advanced Engineering Geology), and EARTHSCI 772 (Hydrogeology) are the core papers. Below, we focus on EARTHSCI 372 and EARTHSCI 771, courses that were specifically impacted due to the timing of lockdowns, and how these papers were adapted from face-to-face delivery to online teaching.

ENGINEERING GEOLOGY COURSES: USUAL DELIVERY

Many of the skills introduced and developed during EARTHSCI 372 are focused on preparing the student for an entry-level role as a graduate engineering geologist. Simply put, 90% of the BSc Earth Science graduates in New Zealand are taken on as engineering geologists to work in the civil engineering sector, and many students attain entry-level roles with a BSc. Therefore, employability is a key function of EARTHSCI 372, providing enough experiential learning opportunities and theory to "scaffold" the BSc Earth Science graduate in their first foray into the workforce. Enrolments range between 30 and 50 students.

EARTHSCI 372 includes basic concepts and principles governing the mechanical behaviour of rock and soil, the properties of rock and rock masses, as well as land instability and ground investigation. Soil behaviour includes phase relationships, effective stress principles, and strength. Rock properties include the failure criterion, description and classification, and strength. Rock mass properties include characteristics of discontinuities, strength and deformability. Land instability and ground investigations are centred around real-world case study material from the conference and journal literature. On a weekly basis, this typically includes 3 × 1-hr lectures, a 2-hr lab session, and 2 hours of reading, which prepares students for the 50% exam. Laboratories include engineering geological mapping, soil particle-size distribution, soil plasticity, core logging, and field-based rock mass classification.

EARTHSCI 771 Advanced Engineering Geology extends skills and knowledge gained at the undergraduate level into the postgraduate domain. Enrolments typically range between 10 and 20 per term. EARTHSCI 771 is more focused on individual learning experiences and is more analytical. For example, at the undergraduate level, the lab provides an opportunity for the students to get an impression of the basic principles of engineering geology. At the postgraduate level, the lab is used as an instrument for advanced level study and for finding solutions to real-life geological engineering problems. The paper comprises a series of 1-day field trips, during which data is collected that form the basis for a 50% report. In addition, the collected data are utilised in five computer-based labs (worth 10% each), focused on the Rocscience suite of software.

ENGINEERING GEOLOGY COURSES: ADAPTATION TO ONLINE TEACHING

Lockdowns in the Faculty of Science mirrored those of the Faculty of Engineering as outlined above. However, enrolment numbers are an order of magnitude fewer in Engineering Geology classes, so the staff–student ratios are lower, and there are also some subtle differences in teaching delivery styles.

Lectures

As with the geomechanics courses, Engineering Geology lecturers heavily utilised the university's LMS, Canvas. For EARTHSCI 372, lectures were given "live" over Zoom, typically with breaks. Given the relatively small class sizes compared with engineering's geomechanics classes, students were encouraged to use both the "reactions" function within Zoom and the "chat" function to stimulate discussion and questions and facilitate engagement. All lectures were also recorded and uploaded to the LMS. Lecture-related readings from academic journal articles, conference proceedings and book chapters were used to augment each lecture. All of these were also provided in Canvas, along with each lecture recording, and a.pdf file of the lecture's PowerPoint presentation. This structuring provided a close parallel to the content and engagement from 2019 and earlier years, and, to a significant degree, simulated the on-campus environment.

In addition, emerging case study material was also used to further encourage engagement. Typically, this was an international landslide event that had been reported in the news media, which was then critically evaluated by the class during a lecture session. A specific example used was the July-August 2021 Parton land instability event in Cumbria (UK), widely covered in a series of BBC news articles (BBC, 2021). It was a particularly useful critical evaluation exercise for the EARTHSCI 372 class. This was because while the village was evacuated in late July 2021 due to apparent tension cracks in the hillside above the village and perceived landslide risk, the villagers were allowed to return on 8 August 2021 as the "tension cracks" were actually caused by drying and shrinkage of the clay-rich soils, not slope failure. Hence, this provided a contemporary, real-world example of the need for thorough site evaluation and monitoring. During the online Zoom class, students provided "rapid-fire" suggestions of site investigation techniques and monitoring approaches that could have been applied to the hillside above Parton. In other lectures, the AGU-hosted Landslide Blog (AGU, 2021), a commentary on worldwide landslides edited by Professor Dave Petley, was used as a vehicle for further engagement.

For the postgraduate-level EARTHSCI 771 Advanced Engineering Geology, the usually scheduled 2-hr lectures, which are traditionally more of a seminar-style format, had to be heavily modified from the usual content for online teaching. This was because the lecture slot contact time was required to bolster the introduction of the computer-based lab sessions (discussed further in Section 8.4.1.2), which form the basis of 50% of the course assessment. Thus, delivering content that allowed the requirements of all the learning outcomes to be attained, proved exigent.

Laboratories

As with the geomechanics papers taught by Civil Engineering, labs proved to be particularly challenging to deliver. In EARTHSCI 372, the five labs are each worth 10%, and represent half of the final assessment grade. Usually, these are either conducted in the School of Environment's Sedimentology Laboratory or Multi-Disciplinary Learning Space (MDLS). The MDLS allows for boxes of core to be laid out across benches for logging, where the students get to touch and feel the rock core, assess texture and mineralogy, as well as assessing defects such as joints and, sometimes, faults. Engineering geological mapping exercises are undertaken using aerial photographs and stereoscopes. In the lab, plasticity is evaluated using liquid and plastic limit tests, and particle-size distribution is evaluated using wet and dry sieving, and a laser particle-size analyser.

With the move to online teaching, some improvisation was necessary. Laboratory introductory videos were provided by the lecturing staff. In some instances, YouTube video clips were provided to demonstrate lab procedures, and spreadsheets of "raw" data from previous years were provided for students to analyse and reflect upon. In addition, live Q&A sessions were provided using Zoom. These Q&A sessions proved especially helpful in addressing student queries and in promoting discussion of results. New readings were introduced so that students could contextualise their answers. The aerial photo lab was moved onto Google Earth, and *.kmz files of the geology were also provided, along with a shaded relief digital elevation model. The core logging lab, where students usually get to apply the New Zealand Geotechnical Guidelines (NZGS, 2005) to core in the Auckland region was the most modified. Instead, core photos and geophysical logs were provided from Carborough Downs underground coal mine in Queensland's Bowen Basin in order to undertake a coal mine roof rating (CMRR) analysis (Mark and Molinda, 2005). The students were able to log the core visually using photos and a measuring tool, and used the corresponding geophysical logs to calculate variation in uniaxial compressive strength (UCS) from a Vp-UCS statistical relationship that was provided to them.

As outlined earlier, lectures for the EARTHSCI 771 Advanced Engineering Geology needed substantial modification in order to provide more background to the labs. This was because the labs are usually scheduled on-campus in a computer lab, which allows the teaching staff to deal with students on a one-on-one basis, as students work their way through the assessed experiments (each of the five labs is worth 10% of the course assessment). All labs are Rocscience-based, utilising the campus-wide Rocscience academic licence. Rocscience was approached by the lecturing staff, and Rocscience very kindly provided each individual student with a temporary licence so that they could download and access the required Rocscience software (Dips, Swedge, RocPlane, Slide2D, RS2) for the labs at home. YouTube and Zoom sessions were then used to facilitate the labs, and submission deadlines were extended accordingly.

Field Trips/Site Visits

Engineering Geology at the University of Auckland has always included site visits and fieldwork to acquire data for later analysis within the course. For EARTHSCI 771, in 2020 and 2021, it was fortunate that scheduled field trips did not clash with the lockdowns, and so these field trips continued unaffected.

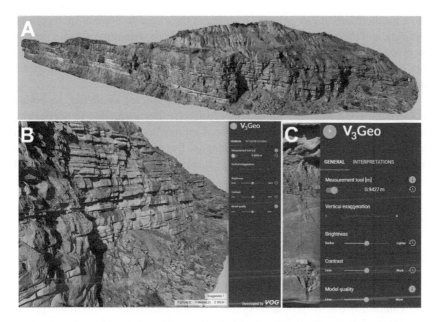

FIGURE 8.2   Ainsa Quarry outcrop visualised in V3Geo: (a) 240 m wide quarry face; (b) tilted and zoomed section; (c) measuring tool used to determine thickness (0.9427 m) of the sandstone layer.

However, for EARTHSCI 372, the 1-day Auckland-based field trip, whereby students visit outcrop, describe and classify it using the Geological Strength Index (GSI), had to be replaced in both 2020 and 2021. Instead, a virtual field trip was designed that paralleled the traditional field trip as closely as possible, based on an online high-resolution digital outcrop model on the V3Geo website (Buckley et al., 2019). The site chosen was a quarry at Ainsa, Spain (Figure 8.2). This 3D outcrop model was chosen because the Neogene-age turbidite sandstones and mudstones in the Ainsa Quarry 3D model very closely resemble the Neogene turbidite sequences in the Auckland region, visible in sea cliffs around Auckland. The V3Geo visualisation allows students to zoom, rotate and measure stratigraphic units on the screen down to several cm, identify joint surfaces, and delineate rock failure mechanisms. Also, the University of Oslo provided a field guide for Ainsa Quarry from their student field trips there over the years. This provided background reading material for the Auckland students.

LESSONS FROM CIVIL ENGINEERING & ENGINEERING GEOLOGY ADAPTATIONS

For qualifications in engineering and geoscience that usually require a range of learning experiences, such as labs and field trips, lockdowns can prove to be particularly challenging (Maraqa et al., 2021). Indeed, it is widely acknowledged that fieldwork is, in particular, a means to integrate concepts taught in the classroom with practical experience, fostering holistic, experiential learning (Boyle et al., 2007; Wilson et al.,

2017). Furthermore, it can enhance learning in the affective domain (perceptions of a subject), and therefore facilitate student learning in the cognitive domain (information processing and learning outcomes (Fuller and France, 2019). To that end, fieldwork and labs provide pedagogic practices that can foster deeper and independent learning among students and are hence important.

While the shift to online teaching posed a significant challenge to both the Geomechanics and Engineering Geology lecturers, they were able to adapt using a combination of tried-and-tested conventional approaches, available digital tools, and the lecturers' individual strengths and strategies. Based on feedback from the end-of-semester summative evaluation tool (SET) surveys, the students found the following to be most helpful in their learning:

- Course materials with lots of information, in-class worked examples, and embedded videos.
- Lecture content broken down into smaller "chunks" and presented via different media (videos, course notes, lecture recordings) helped in understanding the course contents at the students' pace.
- Easy-to-follow demonstrations of lab tests and procedures.
- Online office hours provided opportunities for interaction with staff and peers.
- 3D model visualisation as used in virtual field trips could be used to augment in-person field trips in the future.
- Real-world examples of land instability in the news during the semester of study assisted with engagement in the lecture material and readings.

On the other hand, the online learning challenges that the students highlighted include:

- Lack of opportunity to touch, feel, and play with various types of soils to understand better their properties and how they respond to loading.
- Inability to handle, describe, and classify rocks in the core lab and in the field (though some admitted to doing their own "unofficial" site visits during lockdown).
- Difficulty in doing group projects without the usual interaction required. Some group members had difficulty engaging with their peers due to various non-academic reasons.
- The lack of face-to-face interaction and engagement between lecturers and students (and between students themselves) obstructed the learning process and heavily impacted students' motivations. Some students felt the online experience was isolating.
- Resourcing and time management issues, with some students struggling to find suitable study space, reliable internet connection, suitable digital devices, or worse, had other personal duties to attend to brought about by the pandemic (e.g., taking care of older relatives). An implication was that the workload felt unmanageable.
- Occasionally difficulties with analysis with some of the Rocscience software, such as RS2 at home, in terms of computing time on sometimes old, low-specification home computers.

Based on the lecturers' experience in the remote delivery of Geomechanics and Engineering Geology courses in 2020–2021, some of the recommendations to address future challenges include:

- Increase the capability of teaching staff, especially in the use of online learning and teaching tools, such as H5P, YouTube livestream with sli.do Q&A, etc. For this purpose, there is a need to equip and resource staff to design appropriate and flexible course content that meet varying student needs. Employ more Graduate Teaching Assistants (GTA) to support flexible learning and teaching and engage students and build learning communities.
- In geomechanics in particular, with class sizes as large as 300+ students, satisfying each student's individual needs and predicament would be challenging. It may be better to increase face-to-face delivery of small, interactive sessions, such as a flipped classroom model (but this may have cost implications and require recalibrating teaching workload).
- For field trips and site visits, a combination of in-person and virtual hybrid field experiences might prove the most effective approach for producing a more inclusive and equitable learning environment, especially if LIME software (Buckley et al., 2019) is used to host the V3Geo models.

CONCLUSIONS

To an extent, the "emergency" remote online teaching in the two Geomechanics and two Engineering Geology courses described above can be considered successful. Nevertheless, the student cohorts also highlighted inconsistencies across the range of online teaching adapted across all courses they were attending during the given semester. Hence, lecturers throughout all departments in the university should adopt a consistent and, possibly, best-practice approach, for a better student experience. New "best practices" are emerging globally. For example, in USA universities, COVID-19 prompted a grassroots movement whereby field-course instructors developed virtual field exercises, hosted on the National Association for Geoscience Teachers (NAGT) website "Designing Remote Field Experiences" (NAGT, 2021). Thus, field trips and online labs are pedagogic practices that can be used to foster learning among students. However, traditional field trips and/or exercises that rely on students' abilities to reach field locations or utilise software on rudimentary computers can be exclusionary to some students (Whitmeyer and Dordevic, 2021). Thus, in a post-COVID world, combining in-person site visits and virtual labs and field trips into a hybrid experience may be an effective and attractive approach for students and staff alike. In conclusion, the pandemic has taught us that academics must be innovative in we practice science and facilitate learning.

REFERENCES

AGU (2021). *The landslide blog*. American Geophysical Unione. Retrieved from: https://blogs.agu.org/landslideblog/

BBC (2021). *Parton landslip: evacuees allowed to return home*. 5 August 2021. Retrieved from: https://www.bbc.com/news/uk-england-cumbria-58099737

Boyle A, Maguire S, Martin A, Milsom C, Nash R, Rawlinson S, Turner A, Wurthmann S, Conchie S. (2007). Fieldwork is good: the student perception and the affective domain. *Journal of Geography in Higher Education* 31: 299–317.

Buckley SJ, Ringdal K, Naumann N, Dolva B, Kurz TH, Howell JA, Dewez TJB. (2019). LIME: software for 3-D visualization, interpretation, and communication of virtual geoscience models. *Geosphere* 15(1): 222–235.

Carruthers J. (2020). Sustainability in an era of emerging infectious diseases. *South African Journal of Science* 116(3/4), Art. No. 8043.

Eleanor R. (2020). Kiwis – go home: New Zealand to go into month-long lockdown to fight coronavirus. *The Guardian*, 23 March.

Fuller IC, France D. (2019). Field-based pedagogies for developing learners' independence. In: Walkington H, Hill J, Dyer S (eds), *Handbook for Teaching and Learning in Geography*. Edward Elgar Publishing, Cheltenham.

Maraqa MA, Hamouda M, El-Hassan H, El Dieb A, Hassan AA. (2021). Student perceptions of emergency remote civil engineering pedagogy. *Proceedings of the 2021 IEEE Global Engineering Education Conference (EDUCON)*, April 21–23, Vienna, pp. 583–587.

Mark C, Molinda GM. (2005). The coal mine roof rating (CMRR) – a decade of experience. *International Journal of Coal Geology* 64(1–2): 85–103.

NAGT (2021). *Designing Remote Field Experiences*. National Associate of Geoscience Teachers (USA). Retrieved from: https://nagt.org/nagt/teaching_resources/field/designing_remote_field_experie.html

NZGS (2005). *Guideline for the Field Classification and Description of Soil and Rock for Engineering Purposes*. New Zealand Geotechnical Society, Wellington.

Patrick C-J, Peach D, Pocknee C, Webb F, Fletcher M, Pretto G. (2009). *The WIL [Work Integrated Learning] Report: A National Scoping Study*. Australian Learning and Teaching Council, Brisbane.

Roy C, Metson J. (2021). *Universities of the Future – Adaptation and Transformation to Better Serve the Global Community*. Global Federation of Competitiveness Councils, Washington DC. 6pp.

Whitmeyer SJ, Dordevic M. (2021). Creating virtual geologic mapping exercises in a changing world. *Geosphere* 17(1): 226–243.

Wilson H, Leydon J, Wincentak J. (2017). Fieldwork in geography education: defining or declining? The state of fieldwork in Canadian undergraduate geography programs. *Journal of Geography in Higher Education* 41: 94–105.

Theme 3

Student-Centred Teaching Post-COVID-19

Approaches, Reflections, and Wellbeing

9 Engineering Students' Stress and Mental Health

An Essential Piece in the Retention Puzzle

Amanda Biggs
Griffith University, QLD, Australia

Cynthia Furse
The University of Utah, Utah, USA

CONTENTS

DOI: 10.1201/9781003263180-12

INTRODUCTION

Attracting and retaining engineering students and producing skilled, employable graduates are key challenges for engineering educators and university administrators. In Australia, data for domestic students commencing Bachelor-level engineering degrees indicated that approximately 5% of students drop out of university entirely after completing their first year. In addition, the proportion of students completing a degree (in any field, not necessarily their original engineering degree) varies from 25% after four years from commencement to 75% after nine years from commencement (Australian Council of Engineering Deans, 2019). Strategies to successfully retain engineering students are needed to (a) reduce attrition rates; and (b) promote timely completion of degrees.

The benefits of engineering students' retention and success are not limited to academic institutions but extend to the broader local and global community. A report commissioned by the Royal Academy of Engineering (Cebr, 2016) demonstrated a link between national engineering capacity and economic development; the latter being an important determinant of quality of life for citizens. Worldwide, engineering is an in-demand profession and, in many countries, the need for skilled and qualified engineers exceeds the supply (Geisinger & Raman, 2013; Kaspura, 2019). This gap is likely to widen in the near future as the number of new engineering jobs increases, making the retention and success of engineering students even more critical (Kuley et al., 2015; Torpey, 2018).

Although numerous factors contribute to academic success and retention within engineering programmes, the impact on students' stress and mental health is particularly important. University students, in general, are particularly vulnerable to stress and mental ill-health when the pressures of university coincide with their transition from adolescence to adulthood (Karyotaki et al., 2020). In Australia, for example, a recent study indicated that younger Australians (aged between 16 and 34 years) were more likely to experience high or very high levels of psychological distress compared to other age groups (Australian Bureau of Statistics, 2021). Australian university data indicates that the majority of students (approx. 87%) are aged below 30 years, making mental health a significant issue for university students (Department of Education, 2019). Experiencing stress and mental ill-health undermines students' capacity to learn and fully participate in academic experiences, adversely impacting their academic performance (Heim & Heim, 2021; Karyotaki et al., 2020; Prince, 2015; Storrie et al., 2010). Facilitating learning environments that support engineering students' mental health is, therefore, a crucial step in the process of cultivating their retention and success.

Universities recognise the importance of supporting students' wellbeing, and frequently offer access to professional health services (e.g., counselling) and programmes that assist students to meet their study-related demands (e.g., academic study skills) (Prince, 2015). Despite the availability of these services, many students experiencing distress do not seek the support they need due to the stigma associated with mental ill-health, feelings of hopelessness, and lack of knowledge about accessing services (Storrie et al., 2010). Those students who do seek help often first approach their educators to disclose their stress and mental health concerns (Heim & Heim, 2021). This is unsurprising, as educators represent the "human face" of the university and develop rapport with students within their courses. Educators may find this situation confronting if they feel they lack the knowledge, skills, or confidence to adequately respond to students' stress and mental health concerns (Heim & Heim, 2021).

Along with potentially being students' first point-of-contact, building engineering educators' awareness of stress and mental health is important for three additional reasons. First, educators have a great deal of influence over many of the factors that shape students' experience of stress. Educators' decisions regarding curriculum design, teaching practices, and assessment, as well as their statements and actions, can alter students' self-confidence about their capacity to succeed in engineering (Kuley et al., 2015). These perceptions, in turn, influence students' mental health, engagement, and retention in engineering degrees (Mamaril et al., 2016).

Second, vulnerability to stress and mental ill-health is increased for students studying within high-stress cultures. Studies focusing specifically on engineering students' experience of stress have acknowledged the existence of high-stress and "weed-out" cultures within some engineering programmes (e.g., Geisinger & Raman, 2013; Jensen & Cross, 2021; Kuley et al., 2015). There is a stigma associated with stress in these cultures: when students feel they cannot cope as effectively as their peers, this stigma prevents them from seeking support, compounding their distress and feelings of exclusion (Storrie et al., 2010). Stress, mental health, and attrition are inexorably linked with inclusion; therefore, a clear understanding of how educators influence these issues is particularly important for high-stress disciplines, such as engineering.

Third, research has demonstrated elevated rates of stress and mental health reported by students as a result of the COVID-19 pandemic (Dodd et al., 2021). Responding to these increased rates of student distress placed a considerable burden on educators during an already tumultuous period of change (Heim & Heim, 2021). The COVID-19 pandemic necessitated an abrupt transition from on-campus to online teaching, requiring educators to rapidly master new technology and adapt teaching materials for online delivery. This transition was especially challenging for engineering courses relying on laboratories, hands-on activities, and project-based learning (Espinosa et al., 2021b). Heim and Heim (2021) argued that training and information provided to educators during this time prioritised the adaptation of learning activities to an online environment; in comparison, information about responding effectively to student stress and mental health was scarce, despite the elevated mental health risks posed by the pandemic.

This chapter provides information and strategies for engineering educators who are interested in supporting their students' sense of inclusion and capacity to cope with stress. In the first section (Part 1), an overview of the stress process is presented, explaining how prolonged exposure to stressors gradually erodes students' mental health and engagement with their studies. The second section of the chapter (Part 2) emphasises the critical role of educators as resources who support their students' mental health and inclusion in three key areas of influence: (a) leadership; (b) alignment of demands and resources; and (c) facilitation of a supportive learning environment. The framework presented in this section of the chapter draws on well-established theoretical models that have been successfully applied in multiple educational contexts throughout the world.

While the focus of this chapter is on educators as advocates for students' mental health, three important points need to be emphasised. First, there are numerous factors that contribute to students' stress, mental health, inclusion, and attrition, many of which may be beyond an educator's control. Supporting students' mental health and academic success requires a multilevel approach involving multiple stakeholders, within and external to the university (e.g., Kuley et al., 2015). Second, while students' personal characteristics and circumstances do contribute to stress and attrition, the role of factors external to the students, such as curriculum design, educator support, and the learning environment, have a more substantial impact on these outcomes (Kuley et al., 2015). Third, dealing with students' mental health is seldom explicitly part of an educator's job description (Heim & Heim, 2021) and it is important for educators to familiarise themselves with their institution's processes for referring students to appropriately qualified mental health services and privacy/confidentiality standards. Related to this point, it is important that universities provide staff with adequate training and support to ensure these guidelines and processes are clear for both staff and students (Storrie et al., 2010).

PART 1: WHAT IS STRESS?

Stress, occurring in engineering student populations, can be thought of as a process that is initiated when a student encounters a *stressor*. A stressor is a stimuli, situation, or object that (a) is perceived by the student as harmful, threatening, or challenging; (b) has the potential to overwhelm a student's resources and capacity to cope; and (c) is important to a student's wellbeing, growth, and development (Folkman & Lazarus, 1985). The entire process of perceiving stressors, appraising available resources (such as social support or self-confidence), and initiating coping strategies to resolve the stressor is known as the *transactional stress process* (Folkman & Lazarus, 1985).

There are two primary outcomes of the transactional stress process:

- *Successful resolution*: When a student encounters a stressor and has sufficient resources and capacity to cope with a stressor, the situation is resolved and the student experiences positive emotions (Folkman & Moskowitz, 2004). These positive emotions generate further beneficial outcomes for students, such as intrinsic motivation and engagement, bolstering the student's resource reserves and equipping them to successfully deal with stressors in the future.
- *Unsuccessful resolution*: If a student encounters a stressor, and has insufficient resources or capacity to cope, the stressor remains unresolved, causing the student to experience negative emotions (i.e., distress). Distress associated with unsuccessful resolution motivates a student to continue to employ further coping strategies in an attempt to resolve the stressor. *Strain*

occurs when a student is continually exposed to distress associated with unresolved stressors; over time, strain depletes students' resources, compromising their physical and mental health, engagement, motivation, and performance.

It is important to emphasise that *stressors do not uniformly produce negative consequences; the key issue is whether students possess the necessary resources and capacity to cope with the stressors they do encounter.* If students are unable to effectively meet the challenge imposed by the stressor, strain is likely to occur. However, if students are able to meet the challenge and successfully resolve stressors, then engagement, intrinsic motivation, growth, and development are likely to occur, all of which are important parts of the learning process. In order to effectively support students' stress and mental health, it is important to understand the types of stressors engineering students encounter and the strain outcomes that are likely to occur if stressors remain unresolved.

STRESSORS: THE INITIATORS OF THE STRESS PROCESS

In this section of the chapter, three categories of stressors commonly encountered by engineering students are discussed: (a) high demands, (b) lack of fit, and (c) conflict arising from balancing multiple life roles (i.e., study-life conflict). Three important factors impacting students' experience of stress will also be presented: (a) insufficient recovery, (b) high-stress cultures within engineering, and (c) COVID-19.

Stressors Relating to High Demands

While learning is undoubtedly beneficial and rewarding, it does, at times, place high demands on students: a considerable investment of time and energy is required to successfully complete assessment, master new skills, and acquire new knowledge. Furthermore, these learning-related demands are usually experienced in conjunction with other forms of demands, such as financial pressure (Karyotaki et al., 2020). Demands are aspects of work or study that require sustained physical and psychological effort; prolonged exposure to high demands erodes a student's resources and capacity to cope, subsequently impairing their health and performance (Bakker & Demerouti, 2007).

Undertaking studies in engineering is often perceived to be highly demanding, due to the associated high workload, longer contact hours, and complexity of the content (Schwerin et al., 2021). Many engineering students commence their courses with preconceived concerns that courses may be conceptually complex and dense, requiring a level of knowledge, skill, and commitment beyond their capability (Schwerin et al., 2021). Students' willingness to invest their time, energy, and resources to meet high demands may also be compromised if they believe the course content lacks real-world application. Such beliefs undermine students' intrinsic motivation to study and create negative impressions of the engineering profession as a whole, leading to withdrawal from engineering degrees (Kuley et al., 2015).

Stressors Relating to Lack-of-Fit

Lack-of-fit is another important stressor relevant to engineering students' overall experience of stress. According to the areas of work–life model (Maslach & Leiter, 2008), lack-of-fit (or misfit) between a student and their environment causes distress that motivates the student to resolve the source of misfit. If attempts to resolve the misfit are unsuccessful, strain occurs, compromising health and performance. If balance is maintained, or areas of misfit are successfully resolved, positive outcomes, such as engagement and learning, occur.

Although there are many potential sources of misfit, the areas of work–life model focuses on six specific areas in which a lack of fit between a student and their environment is likely to produce strain and decrease engagement (Maslach, 2003; Maslach & Leiter, 2008):

a) *Demands* that are not well-matched with a student's resources or capacity to cope produce strain (as discussed in the previous section). For example, a student may experience a lack of fit when the course content and teaching practices are not aligned with their learning style and level of academic preparedness (Geisinger & Raman, 2013; Kuley et al., 2015).

b) *Control* involves a student's ability to make important decisions over when and how they manage their demands. An imbalance between control and demands, which may occur when a student has low levels of control over their high demands, is associated with increased strain.

c) *Values* relate to a student's priorities and goals; strain occurs due to a lack of congruence between the values held by a student and those espoused or enacted by others in their environment, their university, and their future chosen profession.

d) *Community* refers to a student's embeddedness in their social networks as well as the quality of their social relationships with significant others, such as their peers and educators. This is linked to their experiences of interpersonal conflict and social support, as well as their sense of belonging at university and within their future chosen profession.

e) *Reward* relates to the financial, social, academic, or internal recognition associated with studying. Lack of fit in relation to rewards can occur for multiple reasons; for example, a mismatch between effort and reward, or when a particular form of recognition is not important to the student.

f) *Fairness* relates to the congruence between a student's sense of equity and justice, and the statements, behaviours, policies, and practices exhibited by others within the university environment.

This suggests that mental health, learning, and engagement are optimised when engineering students experience congruence in each of these six areas. The issue of fit is particularly important for retention in engineering: attrition is reduced when students feel their demands match their abilities, they are accepted by their peers and educators, and they fit within their university's engineering culture (Jensen & Cross, 2021). Experiencing a sense of fit while undertaking engineering studies also reinforces students' perceived future fit within the engineering profession (Kuley et al., 2015).

Stressors Relating to Study-Life Conflict

Traditionally, a typical university student was enrolled full time, entered university directly from high school, and had minimal responsibilities beyond their student role. This is no longer the reality for many students balancing their university demands with responsibilities from other life domains, including family and paid employment (Sprung & Rogers, 2021). Conflict from multiple life roles is a type of stressor that produces strain when demands from different roles compete with one another and overtax students' time, energy, and resources (Kalliath & Brough, 2008). Despite the potential for conflict, involvement in multiple life roles also produces substantial benefits; research in work contexts, for instance, has demonstrated that employees who are encouraged to actively engage in life outside of work exhibit higher levels of work engagement and productivity (Sonnentag et al., 2008). When applied to a university context, respecting students' involvement in multiple life roles, and providing the flexibility students require to manage their study-life conflict, may reduce their stress and enhance their engagement at university.

Statistics indicate (at least in Australia) that engineering is consistently the field of education with the highest rate of school-leaver entry (Australian Council of Engineering Deans, 2019). Based on these statistics, it could be interpreted that balancing multiple life roles may be less of a concern for engineering students who are proportionally more likely to resemble the traditional full-time student. However, the issue of study-life conflict is likely to become increasingly important in the future, and is also necessary for promoting diversity and inclusion within engineering. Students identifying as female are underrepresented in engineering courses; for example, Engineers Australia reported the proportion of female engineering graduates in 2018 was 14.6% (Kaspura, 2020). Lack of work–life balance in STEM professions, including engineering, is a critical barrier to career progression, leading to disproportionately higher rates of attrition from the profession for engineers who identify as female (Dasgupta & Stout, 2014). A greater focus on balancing multiple life roles at all stages of career progression, including at the university level, is an important strategy for attracting and retaining underrepresented groups in engineering (Furse, 2020).

Insufficient Recovery Intensifies Students' Experience of Stress

Stressors most commonly experienced by engineering students, such as high demands, misfit, and study-life conflict, produce strain when students are continuously exposed to them over time. Chronic exposure to stressors and strains impairs health and engagement because it prevents students from recovering from their stress. The relationship between chronic exposure to stress and insufficient recovery resembles a downward spiral: students experiencing high demands typically attempt to meet those demands and maintain their current performance level by exerting additional effort, such as working longer hours (Hockey, 1997). The presence of persistent stressors and strain, insufficient recovery, and increased compensatory efforts to maintain performance erodes resources and seriously compromises students' health and performance over time (Sonnentag et al., 2010). Importantly, exposure to stressful situations may be prolonged if a student worries about a future stressor that may or may not occur (e.g., anticipated poor performance in an upcoming exam) or ruminating about stressors encountered in the past (e.g., previous poor performance; Brosschot et al., 2005). Supporting students' capacity to "switch off" or recover from their study-related demands is an important aspect of supporting their mental health and engagement, which can be challenging when high-stress cultures exist in engineering.

High-Stress Engineering Cultures Intensify Students' Demands and Perceived Lack-of-Fit

All organisations, including academic institutions, possess a culture that reflects the values, beliefs, and expectations for behaviour shared by people within the organisation. Depending on its strength, the culture plays an important role in shaping members' attitudes, behaviours, and sense of belonging (Schein, 2010). An engineering department's culture is influenced by several factors, including the values of its leaders, the broader culture within the academic institution, and the engineering profession itself. Students are socialised into the engineering department's culture through numerous mechanisms, including orientation activities and interactions with educators, alumni, and existing students (Jensen & Cross, 2021).

Students' mental health and sense of belonging may be compromised by educational environments with high-stress or "weed-out" cultures, which are reported to exist within some engineering departments (Geisinger & Raman, 2013; Jensen & Cross, 2021; Kuley et al., 2015). In high-stress and weed-out cultures, being stressed is often equated with being productive, work pressure and competition are promoted, and recovery and active involvement in other life roles may be discouraged (Jensen & Cross, 2021). When high levels of demand are considered to be the norm, students may experience social pressure to "fit" by working at an unsustainably high pace; an inability to cope may be viewed as an indicator that a student does not fit within the engineering department's culture and, by extension, is poorly suited to a career in engineering (Kuley et al., 2015). In high-stress, weed-out cultures, students may be dissuaded from seeking support from their educators and peers, and may struggle to develop a professional engineering identity if they feel unable to perform at the same level as their peers (Jensen & Cross, 2021). Efforts to address engineering students' mental health and sense of belonging are more effective when they are supported by a psychologically safe and inclusive culture.

The Effect of COVID-19 on Students' Experience of Stress

COVID-19 had a significant impact on students' university studies and, subsequently, their experience of stress in several ways. The rapid transition to exclusively online teaching in response to the pandemic required students to quickly adapt to new technologies and ways of interacting with peers and educators online; challenges arising from this transition were exacerbated for students who lacked a suitable work environment or necessary equipment to fully participate in online learning activities during periods of lockdown (Espinosa et al., 2021b). This transition also resulted in disrupted, delayed, and, in some cases, lost learning opportunities for students studying disciplines, such as engineering, where learning activities could not be completely transferred to an online learning environment (Dodd et al., 2021). In many regions, travel restrictions also disrupted the plans of interstate/international students and student exchange programmes.

Beyond their university studies, many working students experienced severe disruptions to their work and capacity to financially support themselves after experiencing reduced hours or redundancy from their work. University students with caring responsibilities were also suddenly required to balance studies with caring and home-schooling demands, when care centres and schools were closed during lockdowns. The impact of the pandemic on the economy and labour market also presented uncertainty about future job prospects, creating more anxiety about students' future success in obtaining work in their chosen professional fields (Dodd et al., 2021). Studies have empirically demonstrated the impact of COVID-19-related stressors on university performance. For example, a study of 787 students studying at Australian universities indicated that approximately 87% reported COVID had a significant impact on their studies (Dodd et al., 2021). COVID-19 has intensified students' experience of stress, making the issue of supporting students' stress and mental health all the more critical.

STRAIN: THE CONSEQUENCES OF UNRESOLVED STRESS

Strain occurs as a consequence of prolonged exposure to the negative emotions evoked by unresolved stressors, including high demands and lack-of-fit. Individual strain responses are often classified as psychological, physiological, and behavioural (Cooper et al., 2001). Psychological strain responses include negative emotions and symptoms of mental ill-health, such as frustration, depression, anxiety, hopelessness, inability to concentrate, forgetfulness, and burnout. Physiological strain responses include headaches, general aches and pains, increased heart rate, upset stomach, dizziness, and muscle tension. Behavioural strain responses include changes in behaviour resulting from stressor exposure, including changes in the amount and frequency of food intake and sleep, engaging in risky behaviour, withdrawal behaviours (from work, study, relationships, leisure activities), aggression, and increased substance use. Continual exposure to stressors and strains may lead to serious physical and psychological illness, potentially compromising students' academic experience, performance, employability, and quality of life (Heim & Heim, 2021; Jensen & Cross, 2021; Karyotaki et al., 2020; Prince, 2015; Storrie et al., 2010). Although research on stress typically focuses on the individual, the consequences of strain discussed above have a deleterious effect on the groups, institutions, and communities in which the individuals are embedded. Strategies to support students' mental health are, therefore, beneficial for universities in addition to their students.

PART 2: PRINCIPLES AND STRATEGIES FOR ADDRESSING STUDENTS' STRESS AND MENTAL HEALTH

In Part 1, the stressors commonly experienced by engineering students were discussed, in addition to strain outcomes that may occur due to prolonged stressor exposure. The severity of these consequences provides a compelling case for prioritising support for students' mental health. Addressing student stress and mental health requires a comprehensive approach that: (a) removes or minimises exposure to unnecessary stressors; (b) provides the resources necessary to assist students to cope with those stressors that cannot be removed entirely; and (c) provides support to students who are already experiencing strain-related consequences of chronic stressor exposure (Murphy, 1996). Understanding the stress process, and the various factors that influence students' experience of stress, is an important initial step in taking action to address the issue. In Part 2 of this chapter, we present a framework of principles and specific strategies educators may draw on to support students' stress and mental health, reflecting the important influence of educators in students' experience of stress at university.

The overarching framework is presented in Figure 9.1, and draws on several theoretical models grounded within the disciplines of organisational and educational psychology (Avolio, 2011; Demerouti et al., 2001; Folkman & Lazarus, 1985; Maslach & Leiter, 2008). The focal point of the model is the stress process (a), described in Part 1 of this chapter. Surrounding the stress process are the three areas of educator influence in students' experience of stress that will be discussed in Part 2 of this chapter: educators' leadership (b), educators' influence over personal and study-related resources (c), and educators' influence over the psychosocial learning environment (d). These three forms of influence can be regarded as multilevel resources that assist students to manage their stress, mental health, and sense of inclusion.

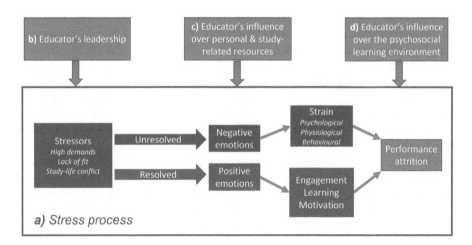

FIGURE 9.1   Theoretical model underpinning principles and strategies for supporting students' wellbeing and engagement.

Educators' Leadership Role in Students' Experience of Stress and Mental Health

Due to their expertise and designated university roles, educators possess formal and informal power that influences students' academic performance and experience of stress. The role of educator, therefore, is inherently a leadership role. Leadership refers to a process of social influence, in which one person (the leader/educator) has the capacity to influence another person (the follower/student) (Kelloway & Barling, 2010). Just as organisational leaders shape employees' experience of work, educators shape students' experience of university by facilitating learning activities, setting clear expectations for assessment, and cultivating a positive learning environment that promotes students' sense of belonging and builds their personal and professional resources (Kuley et al., 2015). Despite educators holding important leadership roles, they are seldom provided with information and support to develop their leadership skills. In this section of the chapter, we will discuss leadership styles that have been demonstrated to be effective in multiple cross-cultural and educational contexts: transformational leadership and authentic leadership.

Transformational leadership is perhaps one of the best known and most influential styles of leadership, and consists of four characteristics:

- *Idealised influence* is demonstrated when educators motivate students by establishing themselves as a role model for students to emulate. Students are motivated to engage in learning when their educators "walk the talk" (i.e., display alignment between their actions and espoused values), demonstrate genuine concern for students' progress, and rely on personal rather than positional sources of power (e.g., drawing on their own expertise and displaying personal characteristics, such as charisma, to inspire students to engage in learning activities (Avolio, 2011; Judge & Piccolo, 2004; Podsakoff et al., 1990; Walumbwa & Wernsing, 2012; Walumbwa et al., 2004)).
- *Inspirational motivation* occurs when educators motivate students by clearly articulating a vision for the future that is appealing to the students, and elevating their interests and expectations. Within engineering courses, students may become quickly demotivated when they perceive the content to be too difficult, tedious, or irrelevant (Warnick & Selvan, 2020). A transformational educator can elevate students' interests and expectations by emphasising the importance of the content and clearly articulating links to real-world applications (i.e., making the content more meaningful); role-modelling enthusiasm for the content; and enhancing students' optimism and self-efficacy (Judge & Piccolo, 2004; Podsakoff et al., 1990). Writing specifically about teaching electromagnetics courses, Warnick and Selvan (2020, p. 28) argued *"teacher enthusiasm is of paramount importance. An enthusiastic teacher can elevate the level of student interest even in subjects perceived to be difficult, practically irrelevant, or both"*.
- *Intellectual stimulation* is displayed when educators motivate students by challenging their existing assumptions, appealing to their intellect, and stimulating their innovation and creativity. Such educators also encourage

students to openly share their ideas and adopt novel approaches to solving problems (Avolio, 2011; Walumbwa & Wernsing, 2012).

- *Individualised consideration* involves educators demonstrating, through their words and actions, respect and concern for students' wellbeing and development, treating them as individuals of value, rather than just another student studying their course. Educators provide individualised consideration by offering support and mentoring to students; facilitating a positive, inclusive learning environment; and encouraging two-way communication, where students feel free to communicate their ideas and concerns (Avolio, 2011; Judge & Piccolo, 2004; Podsakoff et al., 1990).

Integrating transformational leadership behaviours in teaching practices is most effective when combined with authentic leadership and positive aspects of transactional leadership. *Contingent reward* is a specific transactional leadership approach that elicits high performance from followers by clarifying expectations, establishing goals, and distributing rewards in exchange for effort (Avolio, 2011; Walumbwa & Wernsing, 2012). With its focus on setting clear learning goals, performance expectations, and criteria for class participation and assessment, contingent reward assists in maintaining consistent and rigorous standards for academic achievement and integrity, necessary in higher education contexts. Relying on contingent reward leadership alone, however, limits an educator's capacity to inspire, challenge, and develop students (Avolio, 2011); instead, it is useful to consider contingent reward leadership as a necessary foundation on which transformational leadership can be built (Avolio, 2011).

Contingent reward and transformational leadership are most effective when combined with the principles of authentic leadership. There are four characteristics authentic leaders display:

- Educators who are authentic leaders possess *self-awareness* of their own strengths, weaknesses, values, and motivations, and the impact of these factors on their students (Walumbwa, Avolio, Gardner, Wernsing, & Peterson, 2008; Walumbwa & Wernsing, 2012). Seeking feedback and engaging in self-reflection are strategies educators can employ to develop their self-awareness of students' perceptions of their teaching and adapt their teaching practices accordingly (Soares & Lopes, 2020).
- *Relational transparency* exists when educators present an authentic representation of their true selves, by communicating with students in an honest and clear manner (Walumbwa & Wernsing, 2012). Educators who admit to mistakes, say exactly what they mean, and display emotions that genuinely reflect their underlying feelings have high levels of relational transparency (Soares & Lopes, 2020). As a result of their forthright communication, students gain a better understanding of the messages being conveyed by educators and also feel encouraged to exhibit relational transparency themselves, which enhances the learning process (Soares & Lopes, 2020).
- Educators who are authentic leaders acknowledge that their perceptions and decision-making processes are subject to bias, and strive to make fair

and informed decisions through *balanced processing* (Hannah et al., 2011; Walumbwa & Wernsing, 2012). Educators display this facet of authentic leadership by objectively analysing all relevant information and soliciting input from their students prior to making decisions that impact students' experience and performance. This includes decisions about assessment outcomes and academic conduct, as well as those relating to the selection of teaching methods and development of class dynamics (Soares & Lopes, 2020).

- Educators who are authentic leaders possess an *internalised moral perspective*, in which they are able to self-regulate their external behaviour to reflect their internalised values and moral standards. This enables educators to behave in a pro-social and ethical manner, placing shared group goals ahead of their own personal interests (Hannah et al., 2011; Soares & Lopes, 2020; Walumbwa & Wernsing, 2012).

Research has demonstrated the effectiveness of these leadership behaviours cross-culturally and in specific education contexts (Soares & Lopes, 2020; Walumbwa et al., 2004). Transformational and authentic leadership positively influence engineering students' stress, engagement, and performance by increasing students' resources and intrinsic motivation to learn (Bolkan & Goodboy, 2009; Kuley et al., 2015). This positive impact of transformational and authentic leadership on intrinsic motivation is particularly important for retention in engineering; one study demonstrated that a key difference between students who remain in engineering courses, compared to those who leave, was the motivational belief in the intrinsic value of studying engineering (Hensel et al., 2020). Drawing on authentic, transformational leadership to inspire students to recognise the value of engineering, providing a clear line of sight between content and practice, and developing students' self-efficacy, is likely to enhance their intrinsic motivation and engagement in engineering.

ALIGNING RESOURCES WITH ENGINEERING-SPECIFIC STRESSORS

Stress adversely impacts students when they lack the resources they require to cope with their demands. In addition to helping students cope with stress, resources are instrumental in supporting goal achievement, learning, and intrinsic motivation (Bakker & Demerouti, 2007). Resources that are particularly important in the context of student stress and learning include performance feedback, autonomy, flexibility over the scheduling and delivery of learning activities and assessment, social support from educators and peers, and clarity about course expectations. Educators' influence on these resources is both direct and indirect:

- Through curriculum design and teaching practices, educators have a direct influence on students' demands and capacity to access or develop resources.
- Educators can influence the learning environment and display leadership behaviours (e.g., transformational and authentic leadership) that are conducive to the acquisition and development of resources.

When educators positively influence students' capacity to balance their demands and resources, they not only assist students to manage their stress but more effectively meet the needs of increasingly diverse student cohorts.

Growing rates of university participation have led to increased diversity within student populations (Galante et al., 2018); this includes greater participation of students from groups traditionally underrepresented in engineering, in addition to increased diversity in learning styles, responsibilities across multiple life roles, and academic preparedness (Kuley et al., 2015). Academic – particularly mathematic preparedness – is an especially important issue within engineering. As the diversity of engineering student populations increases, so too does the disparity amongst students in their level of prior knowledge of the mathematical principles fundamental to engineering. Students with less mathematics preparation are more susceptible to stress, require more assistance, have lower levels of academic achievement, and are more likely to withdraw from engineering programmes (Geisinger & Raman, 2013; Kuley et al., 2015). It is important to emphasise that academic preparedness disproportionately affects students from groups traditionally underrepresented in engineering, is not an indicator of a student's actual ability, and is beyond the students' control (Kuley et al., 2015). Educators can engage diverse student populations, reduce attrition, and support student mental health and inclusion by designing curriculum and implementing teaching practices that provide important resources, such as greater flexibility; cater to different learning styles; and scaffold the learning of students with varying degrees of academic preparation (Furse & Ziegenfuss, 2021; Kuley et al., 2015). This section of the chapter provides examples of how engineering educators have modified curriculum and teaching practices to increase study-related resources such as performance feedback, flexibility, and social support.

According to Furse and Ziegenfuss (2021), successful course design and delivery contains three essential elements: active learning; multiple modes of delivery; and alignment between learning objectives, course content, and engineering practice. These three elements also create conditions that build students' study-related resources. For example, active learning strategies, such as project- and problem-based learning, align theory and practice, and provide a supportive and realistic context for students to apply their knowledge and develop their skills. Espinosa et al. (2020; 2021a) provide examples of an experiential learning approach implemented in their Advanced Communication Systems course. As part of their assessment, students were required to work in pairs to complete two experimental projects. For each project, students were given six weeks to design, implement, evaluate, and report the outcomes of the experiment. After completing their first project, students were provided with feedback they could utilise to improve their performance in the second project. The students demonstrated performance improvements between the first and second project, and also positively evaluated their experience of the course.

Engineering courses have traditionally been taught face-to-face, and the transition to purely online teaching during the COVID-19 pandemic presented numerous challenges for educators and students. This transition period has highlighted, however, that flexibility for students may be attained through the use of hybrid or blended modes of learning. In their chapter, Espinosa et al. (2021b), outlined novel approaches to teaching electromagnetics courses online, including the use of virtual project demonstrations and take-home laboratory kits that supported students' learning and engagement during the COVID-19 pandemic. Although the return to on-campus classes may be necessary for more applied classes conducted in laboratories, it is worth considering whether some learning activities may be continued online to provide students with greater flexibility and autonomy over their studies. For instance, providing pre-recorded videos for theoretical content can allow students to work through the material at their own pace, and is also an opportunity to promote diversity and inclusion, by incorporating instructors from diverse backgrounds in the videos (Furse & Ziegenfuss, 2020).

The use of flipped or inverted classrooms is another approach that has been successfully employed in engineering courses, where resources, such as readings, activities, and videos are provided to students in advance of their classes, enabling more time to engage in active learning and interaction with educators and peers, either online or face-to-face (Furse & Ziegenfuss, 2021). Furse and Ziegenfuss (2021) discussed their Hybrid-Flexible (HyFlex) model of teaching that is based on the concept of flipped classrooms, combining in-person and online modes of learning. In their HyFlex design, pre-recorded lectures and written materials are provided to students to view online at their preferred time, while scheduled face-to-face sessions are available either in-person or online, with recordings of the sessions being made available afterwards for students to review at their own pace. This model provides students with the flexibility to study at their own pace and move between their preferred modes of learning, as dictated by their life circumstances and learning styles. Furse and Ziegenfuss (2021) provide guidelines and resources for implementing HyFlex teaching in engineering courses.

Finally, Schwerin et al. (2021) discussed a flipped classroom approach that provided scaffolding for students with varying levels of mathematical preparedness and increased students' autonomy over their studies. Students were provided with a series of video bites, which break content into manageable portions, allowing students to progress through content at their own pace. Following the video bites, students attended a weekly lectorial, an online workshop-style class intended to model the application of theory to solving mathematic problems, as well as allowing an opportunity for students to ask questions about the material. The final component of the course consisted of streaming tutorial problems, where students attended an on-campus workshop to work through problems classified according to three levels of difficulty: starter, intermediate, and advanced. Students were advised their final exam would consist of problems at an intermediate level of difficulty. This strategy provided a more tailored approach to learning that respected the autonomy of learners and met the needs of students with varying levels of mathematical ability. From the students' perspective, it also provided an opportunity for immediate performance feedback regarding their progress in the course: students were advised to seek assistance if they were struggling with starter-level problems. From the educator's perspective, it provided valuable feedback on the concepts that could be moved through quickly versus the concepts needing further exploration.

These examples demonstrate some of the ways in which curriculum design and teaching practices can provide students with important study-related resources, such as flexibility, support, autonomy, and feedback, which help them cope with their demands. When students have access to well-resourced learning environments and learning activities with clear application to real-world engineering problems, they also build personal resources that support their learning and resistance to stress (Hobfoll et al., 2003). Substantial research has demonstrated that personal resources, such as self-efficacy and resilience, are linked to higher levels of academic performance, satisfaction, and retention. For example, engineering students' confidence in their ability to succeed has been positively associated with course grades for engineering students completing an algebra course (Hsieh et al., 2012) and academic achievement, after controlling for students' prior achievement (Mamaril et al., 2016). A review of research on engineering students' attrition also demonstrated that lower self-efficacy is related to higher attrition, particularly for students from underrepresented groups, despite the students with lower self-efficacy not differing from their peers in relation to their actual levels of academic ability (Kuley et al., 2015). Personal resources, such as self-efficacy, are predominantly shaped by external factors, such as curriculum design, effective leadership, and supportive, resourceful learning environments (Kuley et al., 2015; Zafft et al., 2015). The examples presented in this chapter demonstrate strategies for modifying curriculum and teaching practices that provide important study-related resources (e.g., flexibility) and personal resources (e.g., self-efficacy) that assist students to manage their stress, cater to different learning styles and life circumstances, and engage student populations, ultimately alleviating attrition and mental ill-health (Kuley et al., 2015). They also send a clear message that students with diverse learning styles and life circumstances can succeed in engineering, counteracting the notion endemic within high-stress cultures that engineering is only suited to certain "types" of people (Jensen & Cross, 2021; Kuley et al., 2015).

Supporting Students' Stress and Retention by Creating Psychologically Safe and Inclusive Learning Cultures

Strategies for supporting students' mental health and retention are more likely to succeed when engineering departments promote psychologically safe and inclusive learning cultures. In their review of factors contributing to engineering student attrition, Kuley et al. (2015) found that curriculum design and the institutional culture experienced by students were the two most influential determinants of attrition from engineering. Furthermore, while students who withdraw from engineering are academically similar to those who remain, they tend to report more negative perceptions of student-educator interactions and a pervasive feeling their institution does not want them to succeed (Kuley et al., 2015). It has long been recognised that learning is an inherently social process, and interpersonal interactions between learners and educators, in addition to the broader learning environment, have a significant influence over students' learning and their experience of stress and inclusion (Bandura, 1997; Kuley et al., 2015; Soares & Lopes, 2020). As leaders, educators have the capacity to positively influence their students' learning environment by facilitating a culture that promotes psychological safety and sense of belonging.

PROMOTING PSYCHOLOGICALLY SAFE LEARNING ENVIRONMENTS

Psychosocial safety is an important aspect of the learning environment, and refers to learners' perceptions of the consequences associated with taking risks in educational contexts (Edmondson & Lei, 2014). Learning is optimised when *students feel free to take interpersonal risk*, such as asking questions, expressing opinions, sharing information, testing ideas, and seeking feedback, *without fear of negative consequences*, such as being ridiculed or viewed as incompetent by their educators and peers (Edmondson & Lei, 2014; Soares & Lopes, 2020). Psychological safety is linked to positive outcomes such as increased engagement, learning, performance, proactivity, and creativity, and reduced stress (Edmondson & Lei, 2014). Positive, psychologically safe learning environments are especially important when learning and performance occur during periods of uncertainty, which is especially relevant for teaching during the COVID-19 pandemic (Edmondson & Lei, 2014).

Although there are multiple factors contributing to students' experience of psychological safety at university, educators play an important role as leaders within their courses. In their review, Edmondson & Lei (2014) indicated that transformational and authentic leadership behaviours are important antecedents of psychological safety: educators who display these behaviours in their interactions with students are likely to facilitate an environment where students feel free to take interpersonal risks. Through their speech and actions, educators also convey messages about whether students will be supported or opposed for taking interpersonal risks; as such, it is important that educators clearly and consistently encourage students to take interpersonal risks and support them when they do (Edmondson & Lei, 2014). Finally, psychologically safe environments encourage students to take interpersonal risks that not only benefit their learning outcomes but also support their mental health (i.e., seeking support to cope with stress). Students feel free to communicate their concerns about stress and mental health in psychologically safe environments where their wellbeing is prioritised and when leaders (both educators and university administrators) demonstrate support and concern for students, and are committed to addressing the issue of student mental health (Hall et al., 2010).

CREATING A LEARNING ENVIRONMENT THAT SUPPORTS STUDENTS' SENSE OF BELONGING

A sense of belonging is developed in positive, psychologically safe learning environments, and occurs when students feel accepted by, and connected to, their peers, educators, and institutions (Ahn & Davis, 2020). Research has demonstrated that a sense of belonging in STEM is a critical factor determining student retention in STEM disciplines (Rattan et al., 2018). Furthermore, engineering students' sense of belonging, inclusion, and acceptance are critical factors contributing to their retention, performance, and mental health (Hensel et al., 2020; Kuley et al., 2015). Sense of belonging has such a profound impact on students' academic, social, and psychological outcomes because it satisfies the fundamental human need for relatedness, which drives intrinsic motivation and acts as a buffer against stress (Ryan & Deci, 2000).

When universities create psychologically safe environments that cultivate students' sense of belonging, students are intrinsically motivated to invest the effort needed to successfully complete their studies, reducing attrition from engineering degrees (Kuley et al., 2015).

Research has demonstrated the important role educators play in creating learning environments that nurture students' sense of belonging. For example, Freeman et al. (2007) examined the antecedents and consequences of students' sense of belonging at two levels: class-level and institution-level. Educator characteristics that were significantly associated with class-level sense of belonging included students' perceptions that their educators: (a) were friendly, warm, enthusiastic, and helpful; (b) actively encouraged students' participation in class; and (c) were organised and well-prepared for class. Other factors, such as students' gender, high school GPA, and number of out-of-class contacts with the instructor, were not significant predictors. Their study also found that students who felt they belonged within a class perceived the content to be more important and relevant; were more intrinsically motivated to study; and held more positive beliefs about their own self-efficacy within the class. Finally, Freeman et al. (2007) found that perceived social acceptance by peers and educators, and level of caring from educators, were related to students' sense of university-level belonging.

Establishing a sense of belonging appears to be especially critical in the retention of students from underrepresented groups within STEM disciplines, including engineering. Rattan et al. (2018), for instance, conducted six studies that examined how sense of belonging within STEM is influenced by students' perceptions of their educators' beliefs about the universality of scientific aptitude; that is, whether students perceive their educators believe that most individuals have the capacity to succeed in STEM (defined by the researchers as a *universal metatheory*) or that only certain individuals naturally possess the scientific aptitude required to succeed in STEM (defined by the researchers as a *non-universal metatheory*). Students' perceptions of educators' beliefs were the focus of these studies, as students, especially those from underrepresented groups, view educators as gatekeepers who encourage or discourage their pursuit of STEM careers, both academically and professionally. The series of six studies by Rattan et al. (2018) demonstrated that universal meta-theories are positively related to students' sense of belonging. For example, in Study 3, universal meta-theories (measured in Week 3 of the course) positively predicted the final course grade (measured in Week 10), through the mediating effect of sense of belonging to the course (also measured in Week 3). This significant effect occurred even after controlling for midterm exam performance. Overall, the studies conducted by Rattan et al. (2018) demonstrated that students, particularly those from underrepresented groups, are more likely to feel they belong in STEM when they perceive that their educators believe STEM aptitude is universal. It is important to emphasise that these studies examined *students' perceptions of their educators' beliefs*, rather than the *educators' actual beliefs* about the universality of STEM aptitude. It is entirely possible that students' perceptions do not accurately reflect their educator's actual beliefs and, in some instances, educators may unintentionally signal, through their words or actions, that not all people are suited to STEM careers. Therefore, educators wishing to encourage and engage all students, especially those from underrepresented groups, should aim to clearly articulate their belief in their students' capacity to succeed, and ensure that their behaviours, practices, and enactment of policies are aligned with these beliefs.

CONCLUSIONS

Students' experience of stress and inclusion are inexorably linked, and both significantly impact students' health and academic success. The need to address these issues has been recently highlighted, with the intensification of stress arising from the COVID-19 pandemic. This chapter provided information and strategies for engineering educators wishing to take steps towards creating a positive, supportive learning environment inclusive of all students. Part 1 provided an overview of the stress process, explaining how continual exposure to high demands, study-life conflict, and lack-of-fit gradually erodes the resources needed for students to successfully cope with stress. The influence of insufficient recovery, high-stress cultures, and COVID-19 was also discussed. An awareness of this process is an important first step in becoming an advocate for students' mental health and inclusion.

While there are multiple, complex factors that influence students' stress and inclusion, there are strategies educators can incorporate in their teaching practices to engage their students, help them manage their stress, and build their feelings of inclusion. While the uncertainty and change occurring as a result of the COVID-19 pandemic created numerous challenges for educators, it also offered an opportunity to reflect on how curriculum and teaching practices can be designed or modified to provide students with a more supportive, inclusive learning environment. The following strategies were discussed in Part 2:

- Embracing the leadership component of an educator's role, role-modelling enthusiasm for engineering, and clearly articulating the importance and real-world impact of the content being taught.
- Demonstrating self-awareness, authenticity, and transparency when communicating with students, encouraging them to feel safe to reciprocate in an equally transparent manner.
- Communicating an awareness of students as individuals with unique values, goals, and needs.
- Considering how modifications to curriculum design and teaching practices can provide students with access to the study-related and personal resources that help them cope with stress and promote learning and development.
- Promoting psychologically safe learning environments, where students feel free to take interpersonal risks conducive to learning and that foster a sense of belonging for all students.
- Clearly and consistently articulating your belief in your students' capacity to succeed in engineering, as well as the importance of their wellbeing as a priority.
- Demonstrating sensitivity to the ongoing challenges relating to the COVID-19 pandemic that will continue to intensify students' experience of stress in the foreseeable future.

- Familiarising yourself with your institution's processes and systems for referring students to appropriately qualified health services, guidelines relating to student privacy/confidentiality (especially important for instances when duty of care overrides confidentiality), and support services available to provide you with advice and support when dealing with students experiencing stress.

These strategies facilitate the kind of learning environment that enables students to access or develop the resources they need to succeed in engineering, and develop the sense that engineering is a good fit for them (Kuley et al., 2015). Given the essential role engineering plays in economic development and quality of life, and the ever-growing need for novel engineering solutions to societal challenges, the retention and success of diversely talented engineering graduates is something we should all care about.

REFERENCES

Ahn, M. Y., & Davis, H. (2020). Students' sense of belonging and their socio-economic status in higher education: A quantitative approach. *Teaching in Higher Education*. doi:10. 1080/13562517.2020.1778664

Australian Bureau of Statistics. (2021). First insights from the National Study of Mental Health and Wellbeing, 2020–21. Retrieved from https://www.abs.gov.au/articles/first-insights-national-study-mental-health-and-wellbeing-2020-21

Australian Council of Engineering Deans. (2019). *Australian Engineering Education Student, Graduate, and Staff Data and Performance Trends*. Retrieved from https://www.aced. edu.au/downloads/ACED%20Engineering%20Statistics%20March%202019.pdf

Avolio, B. J. (2011). *Full range leadership development*. (2nd ed.). Thousand Oaks, CA: Sage Publications.

Bakker, A. B., & Demerouti, E. (2007). The job demands-resources model: State of the art. *Journal of Managerial Psychology*, 22(3), 309–328. doi:10.1108/02683940710733115

Bandura, A. (1997). *Self-efficacy: The exercise of control*. New York: Freeman Press.

Bolkan, S., & Goodboy, A. K. (2009). Transformational leadership in the classroom: Fostering student learning, student participation, and teacher credibility. *Journal of Instructional Psychology*, 36(4), 296–306.

Brosschot, J. F., Pieper, S., & Thayer, J. F. (2005). Expanding stress theory: Prolonged activation and perseverative cognition. *Psychoneuroendocrinology*, 30, 1043–1049.

Cebr. (2016). *Engineering and Economic Growth: A Global View*. Retrieved from https://www. raeng.org.uk/publications/reports/engineering-and-economic-growth-a-global-view

Cooper, C. L., Dewe, P., & O'Driscoll, M. P. (2001). *Organizational stress: A review and critique of theory, research, and applications*. Thousand Oaks, CA: Sage Publications, Inc.

Dasgupta, N., & Stout, J. G. (2014). Girls and women in science, technology, engineering, and mathematics: STEMing the tide and broadening participation in STEM careers. *Policy Insights from the Behavioral and Brain Sciences*, 1(1), 21–29. doi:10.1177/2372732214549471

Demerouti, E., Bakker, A. B., Nachreiner, F., & Schaufeli, W. B. (2001). The job demands-resources model of burnout. *Journal of Applied Psychology*, 86(3), 499–512. doi:10.1037//0021-9010.86.3.499

Department of Education, Skills and Employment. (2019). *Higher Education Statistics*. Retrieved from https://www.dese.gov.au/higher-education-statistics/resources/2019-section-2-all-students

Dodd, R. H., Dadaczynski, K., Okan, O., McCaffery, K. J., & Pickles, K. (2021). Psychological wellbeing and academic experience of university students in Australia during COVID-19. *International Journal of Environmental Research and Public Health, 18*(3), 866. doi:10.3390/ijerph18030866

Edmondson, A. C., & Lei, Z. (2014). Psychological safety: The history, renaissance, and future of an interpersonal construct. *Annual Review of Organizational Psychology and Organizational Behavior, 1*, 23–42. doi:10.1146/annurev-orgpsych-031413-091305

Espinosa, H. G., Fickenscher, T., Littman, N., & Thiel, D. V. (2020). *Teaching wireless communications courses: An experiential learning approach.* Paper presented at the *2020 14th European Conference on Antennas and Propogation (EuCAP).*

Espinosa, H. G., James, J., Littman, N., Fickenscher, T., & Thiel, D. V. (2021a). An experiential learning approach in electromagenetics education. In K. T. Selvan & K. F. Warnick (Eds.), *Teaching electromagnetics: Innovative approaches and pedagogical strategies* (pp. 21–47). CRC Press.

Espinosa, H. G., Khankhoje, U., Furse, C., Sevgi, L., & Rodriguez, B. S. (2021b). Learning and teachin in a time of pandemic. In K. T. Selvan & K. F. Warnick (Eds.), *Teaching electromagnetics: Innovative approaches and pedagological strategies.* (pp. 219–237). CRC Press.

Folkman, S., & Lazarus, R. S. (1985). If it changes it must be a process: Study of emotion and coping during three stages of a college examination. *Journal of Personality and Social Psychology, 48*(1), 150–170.

Folkman, S., & Moskowitz, J. T. (2004). Coping pitfalls and promise. *Annual Review of Psychology, 55*, 745–774.

Freeman, T. M., Anderman, L. H., & Jensen, J. M. (2007). Sense of belonging in college freshmen at the classroom and campus levels. *The Journal of Experimental Education, 75*(3), 203–220. doi:10.3200/JEXE.75.3.203-220

Furse, C. (2020). How to be a great advocate for women in engineering. *IEEE Antennas & Propogation Magazine, December 2020*, 98–103. doi:10.1109/MAP.2020.3027265

Furse, C., & Ziegenfuss, D. (2020). A busy professor's guide to sanely flipping your classroom. *IEEE Antennas & Propogation Magazine, April 2020*, 31–42. doi:10.1109/MAP.2020.2969241

Furse, C., & Ziegenfuss, D. (2021). HyFlex Flipping: Combining in-person and online teaching for the flexible generation. In K. T. Selvan & K. F. Warnick (Eds.), *Teaching electromagnetics: Innovative approaches and pedagogical strategies* (pp. 201–218). CRC Press.

Galante, J., Dufour, G., Vainre, M., Wagner, A. P., Stochl, J., Benton, A., … Jones, P. B. (2018). A mindfulness-based intervention to increase resilience to stress in university students (the Mindful Student Study): A pragmatic randomised controlled trial. *The Lancet, 3*(2), 72–81.

Geisinger, B. N., & Raman, D. R. (2013). Why they leave: Understanding student attrition from engineering majors. *International Journal of Engineering Education, 29*(4), 914–925.

Hall, G. B., Dollard, M. F., & Coward, J. (2010). Psychosocial safety climate: Development of the PSC-12. *International Journal of Stress Management, 17*(4), 353–383. doi:10.1037/a0021320

Hannah, S. T., Avolio, B. J., & Walumbwa, F. O. (2011). Relationships between authentic leadership, moral courage, and ethical and pro-social behaviors. *Business Ethics Quarterly, 21*(4), 555–578.

Heim, C., & Heim, C. (2021). Facilitating a supportive learning experience: The lecturer's role in addressing mental health issues of university students during COVID-19. *Journal of University Teaching & Learning Practice, 18*(6), 69–81. doi:10.53761/1.18.6.06

Hensel, R. A. M., Dygert, J., & Morris, M. L. (2020). *Understanding student retention in engineering.* Paper presented at the *ASEE's Virtual Conference: At Home with Engineering Education.*

Hobfoll, S. E., Johnson, R. J., Ennis, N., & Jackson, A. P. (2003). Resource loss, resource gain, and emotional outcomes among inner city women. *Journal of Personality and Social Psychology*, *84*(3), 632–643. doi:10.1037/0022-3514.84.3.632

Hockey, G. R. J. (1997). Compensatory control in the regulation of human performance under stress and high workload: A cognitive-energetical framework. *Biological Psychology*, *45*, 73–93.

Hsieh, P., Sullivan, J. R., Sass, D. A., & Guerra, N. S. (2012). Undergraduate engineering students' beliefs, coping strategies, and academic performance: An evaluation of theoretical models. *The Journal of Experimental Education*, *80*(2), 196–218. doi:10.1080/00220973.2011.596853

Jensen, K. J., & Cross, K. J. (2021). Engineering stress culture: Relationships among mental health, engineering identity, and sense of inclusion. *Journal of Engineering Education*, *110*(2), 371–392. doi:10.1002/jee.20391

Judge, T. A., & Piccolo, R. F. (2004). Transformational and transactional leadership: A meta-analytic test of their relative validity. *Journal of Applied Psychology*, *89*(5), 755–768. doi:10.1037/0021-9010.89.5.755

Kalliath, T., & Brough, P. (2008). Work–life balance: A review of the meaning of the balance construct. *Journal of Management and Organization*, *14*(3), 323–327. doi:10.5172/jmo.837.14.3.323

Karyotaki, E., Cuijpers, P., Albor, Y., Alonso, J., Auerbach, R. P., Bantjes, J., … Kessler, R. C. (2020). Sources of stress and their associations eith mental disorders among college students: Results of the World Health Organization World Mental Health Surveys International College Student Initiative. *Frontiers in Psychology*, *11*. doi:10.3389/fpsyg.2020.01759

Kaspura, A. (2019). *The Engineering Profession: A Statistical Overview*. Retrieved from ACT, Australia: https://www.aced.edu.au/downloads/The%20Engineering%20Profession,%20A%20Statistical%20Overview,%2014th%20edition%20-%2020190613.pdf

Kaspura, A. (2020). *Australia's next generation of engineers: University statistics for engineering*. Retrieved from ACT, Australia: https://www.engineersaustralia.org.au/sites/default/files/Higher_education_statistics_2020.pdf

Kelloway, E. K., & Barling, J. (2010). Leadership development as an intervention in occupational health psychology. *Work & Stress*, *24*(3), 260–279. doi:10.1080/02678373.2010.518441

Kuley, E. A., Maw, S., & Fonstad, T. (2015). *Engineering student retention and attrition literature review*. Paper presented at the *2015 Canadian Engineering Education Association Conference*, Ontario.

Mamaril, N. A., Usher, E. L., Li, C. R., Economy, D. R., & Kennedy, M. S. (2016). Measuring undergraduate students' engineering self-efficacy: A validation study. *Journal of Engineering Education*, *105*(2), 366–395. doi:10.1002/jee.20121

Maslach, C. (2003). Job burnout: New directions in research and intervention. *Current Directions in Psychological Science*, *12*(5), 189–192. doi:10.1111/1467-8721.01258

Maslach, C., & Leiter, M. P. (2008). Early predictors of job burnout and engagement. *Journal of Applied Psychology*, *93*(3), 498–512.

Murphy, L. R. (1996). Stress management in work settings: A critical review of the health effects. *American Journal of Health Promotion*, *11*(2), 112–135.

Podsakoff, P. M., MacKenzie, S. B., Moorman, S. B., & Fetter, R. (1990). Transformational leader behaviours and their effects on followers' trust in leader, satisfaction, and organizational citizenship behaviors. *Leadership Quarterly*, *1*(2), 107–142. doi:10.1016/1048-9843(90)90009-7

Prince, J. P. (2015). University student counseling and mental health in the United States: Trends and challenges. *Metal Health & Prevention*, *3*(1–2), 5–10. doi:10.1016/j.mhp.2015.03.001

Rattan, A., Savani, K., Komarraju, M., Morrison, M. M., Boggs, C., & Ambady, N. (2018). Meta-lay theories of scientific potential drive underrepresented students' sense of belonging to science, technology, engineering and mathematics (STEM). *Journal of Personality and Social Psychology, 115*(1), 54–75. doi:10.1037/pspi0000130

Ryan, R. M., & Deci, E. L. (2000). Self-determination theory and the facilitation of intrinsic motivation, social development and wellbeing. *American Psychologist, 55*. doi:10.1037/0003-066x.55.1.68

Schein, E. H. (2010). *Organizational culture and leadership*. (4th ed.). San Francisco, CA: Jossey-Bass.

Schwerin, B., Espinosa, H. G., Gratchev, I., & Lohmann, G. (2021). *Enhancing maths teaching resources: Topic videos and tutorial streaming development*. Paper presented at the *Research in Engineering Education Symposium & Australasian Association for Engineering Education Conference*, Perth, Australia.

Soares, A. E., & Lopes, M. P. (2020). Are your students safe to learn? The role of lecturer's authentic leadership in the creation of psychologically safe environments and their impact on academic performance. *Active Learning in Higher Education, 21*(1), 65–78. doi:10.1177/1469787417742023

Sonnentag, S., Mojza, E. J., Binnewies, C., & Scholl, A. (2008). Being engaged at work and detached at home: A week-level study on work engagement, psychological detachment, and affect. *Work & Stress, 22*(3), 257–276.

Sonnentag, S., Binnewies, C., & Mojza, E. J. (2010). Staying well and engaged when demands are high: The role of psychological detachment. *Journal of Applied Psychology, 95*(5), 965–976. doi:10.1037/a0020032

Sprung, J. M., & Rogers, A. (2021). Work–life balance as a predictor of college student anxiety and depression. *Journal of American College Health, 69*(7), 775–782. doi:10.1080/07448481.2019.1706540

Storrie, K., Ahern, K., & Tuckett, A. (2010). A systematic review: Students with mental health problems-A growing problem. *International journal of nursing practice, 16*(1), 1–6. doi:10.1111/j.1440-172X.2009.01813.x

Torpey, E. (2018). Engineers: Employment, pay, and outlook. *Career Outlook*. Retrieved from https://www.bls.gov/careeroutlook/2018/article/engineers.htm

Walumbwa, F. O., & Wernsing, T. (2012). From transactional and transformational leadership to authentic leadership. In M. G. Rumsey (Ed.), *The Oxford handbook of leadership*. New York: Oxford University Press. doi:10.1093/oxfordhb/9780195398793.013.0023

Walumbwa, F. O., Wu, C., & Ojode, L. A. (2004). Gender and instructional outcomes: The mediating role of leadership style. *Journal of Management Development, 23*(2), 124–140. doi:10.1108/02621710410517229

Walumbwa, F. O., Avolio, B. J., Gardner, W. L., Wernsing, T. S., & Peterson, S. J. (2008). Authentic leadership: Development and validation of a theory-based measure. *Journal of Management, 34*(1), 89–126. doi:10.1177/0149206307308913

Warnick, K. F., & Selvan, K. T. (2020). Teaching and learning electromagnetics in 2020: Issues, trends, opportunities, and ideas for developing courses. *IEEE Antennas & Propagation Magazine, 63*, 24–30.

Zafft, C., McElravy, L. J., & Curtis, E. (2015). *PsyCap to promote and assess professional skils for undergraduate engineering students*. Paper presented at the *7th First Year Engineering Experience Conference*, Roanoke, VA.

10 Evaluating the Impact of COVID-19 Pandemic on Students' Learning Experiences

Sanam Aghdamy, Savindi Caldera, Cheryl Desha, and Mahan Mohammadi
Griffith University, QLD, Australia

Amin Vakili
Amicus Medical Chambers, Brisbane, Australia

CONTENTS

INTRODUCTION

With the unprecedented circumstances of the COVID-19 pandemic, there were urgent calls for higher education institutions (HEIs) to shift from face-to-face to online delivery of coursework (Caldera et al., 2022; Taylor-Guy & Chase, 2020). Online teaching encapsulates all types of teaching that occur through an electronic network, including fully online or a combination of online and face-to-face (i.e., blended) activities. While some HEIs have integrated online teaching elements in a format of asynchronous and static types (e.g., digital platforms to share teaching materials, lecture recordings, and discussions) for some time, this pandemic has urged an immediate shift from face-to-face to online course delivery, specifically, through live streaming (Xiong et al., 2020). This rapid transition from face-to-face to online delivery midway through delivering a course was coined in 2020 as Emergency Remote Teaching (ERT). ERT describes the temporary elevated online learning and teaching environment, which is expected to return to its original face-to-face format when the crisis subsides (Hodges et al., 2020).

Creating an effective online learning environment can be challenging, and demands the purposeful design of learning activities for more effective student engagement. This requires a certain level of Pedagogical Content Knowledge (PCK) to create a distinctive learning environment with digital technologies. Three types of presence (social, cognitive, and facilitatory) were highlighted as significant elements for improved student engagement and online presence (Rapanta et al., 2020). Maintaining a slow voice, sharing resources before the class begins, allowing student feedback, offering flexible learning and teaching policies, and recording online lectures were identified as key strategies for enhancing student learning (Mahmood, 2021). For this experience to be even more effective, the role of an online educator needs to be broadened into several different forms, including "professional", "pedagogical", "social", "evaluator", "administrator", "technologist", "advisor", and "researcher" (Bawane & Spector, 2009). This means that the transition to online teaching may require moving beyond the traditional roles and learning and teaching activities (ibid.).

However, there have been concerns regarding the effectiveness and efficacy of the online delivery of course content, especially in terms of monitoring performance (Alrefaie et al., 2020). A clear monitoring portfolio with appropriate indicators is critical to ensure an evidence-based and purposeful evaluation of the online learning experience (ibid.). An Australian study revealed that COVID-19 has had a significant impact on student attitudes towards digital infrastructure, with digital service capabilities a key criterion in selecting universities. Within this context, 31% of respondents in 2020 said they would switch institutions for a better technological experience, compared to 17% in 2019 (TechnologyOne, 2020). Gibb's reflective cycle was used as an established theoretical foundation to analyse student experience and reflections on both positive and negative experiences (Hill & Fitzgerald, 2020). Students shared their lived experiences, including challenges such as reduced engagement and disrupted learning opportunities due to connectivity and lack of internet infrastructure, financial difficulties, limited supporting devices, other competing household duties, and electronic accessibility (Dhawan, 2020; Henaku, 2020).

While there is a large body of literature on "online teaching", there is limited research on the lived experiences of students and the impacts on their wellbeing and academic achievement during this ERT period. Therefore, this research enquires into the impact of COVID-19 on students' lived experiences, including learning experiences, wellbeing, and academic performance. This study also uncovered opportunities for more effective forward planning, to create resilient student learning environments amidst current and future disruptions.

METHODS

This study was conducted at the School of Engineering and Built Environment, Griffith University, in consultation with a psychiatric consultant from Amicus Medical Chambers, a psychiatric clinic in Queensland, Australia. The Ethics clearance for conducting this research was obtained from the Research Ethics and Integrity division of the Griffith University Research Office (Reference ID: 2020/420). It involved conducting three rounds of 15-minute surveys over 1.5 years to investigate the impact of COVID-19 at different phases on students' lived experiences. A survey is an appropriate data collection method that can obtain both quantitative and qualitative information using well-planned questionnaires. This approach is widely used by researchers within the engineering education domain (Abowitz & Toole, 2010; Ponto, 2015). An online questionnaire was used in this study as it is an efficient and flexible tool that ensures participant confidentiality. It also implements the most common delivery method; participants are familiar with this approach and are more likely to respond. The survey was conducted using LimeSurvey (n.d.) and the data were analysed using IBM SPSS software (IBM Corp., 2016). The survey enabled the researchers to elicit knowledge about barriers and opportunities for effective online curriculum delivery during disruptions (Creswell, 2013; Mrug, 2010).

SURVEY DESIGN AND DATA COLLECTION

The first round of the survey was performed at the end of the first academic trimester in May 2020, during the early stage of COVID-19 pandemic lockdown in Australia, wherein urgent and disruptive transformations of learning and teaching took place (i.e., Phase I). The second round of the survey was performed at the end of the second academic trimester in October 2020, when students and teachers were settling into the new online learning and teaching environment (i.e., Phase II). The third round of survey was conducted at the end of the first trimester in May 2021, wherein COVID-19 restrictions were easing, resulting in some on-campus/face-to-face activities (i.e., Phase III). This chapter reports the results of the first two rounds of survey.

The research included undergraduate and postgraduate students from Civil Engineering, Electronics and Electrical Engineering, Environmental Engineering, and Engineering Project Management programmes. Table 10.1 summarises the nominated Engineering courses in each trimester. In this table, *NA* and *GC* represent the two different campuses of Nathan and Gold Coast, respectively. Since the survey was conducted over two consecutive trimesters, the courses have been selected to ensure that the majority of participants remain the same throughout these trimesters. In order to create the ability to connect students' data while preserving their anonymity, participants were asked to create, and then always use, the same unique identifier. To minimise the risk of forgetting the identifier, students were asked to use the name of the first street in which they had lived, followed by the colour of their eyes (e.g., "queenbrown"). Students also had a choice of using a different identifier if they wished to.

TABLE 10.1

A Summary of the Nominated Engineering Courses in Each Trimester

Trimester 1, 2020	Trimester 2, 2020
■ Mechanics of Materials I (2101ENG) -NA, GC	■ Structural Design (2103ENG) – NA, GC
■ Fluid Mechanics and Hydraulics (2002ENG) – NA, GC	■ Engineering Thermodynamic (2201ENG) – NA
■ Structural Analysis (3101ENG) – NA, GC	■ Communication Systems and Circuits (2304ENG) – NA
■ Control Systems (3304ENG) – NA, GC	■ Concrete Structures (3103ENG) – NA, GC
■ Structural Analysis (7310ENG) – GC	■ Renewable Energy Systems (3203ENG) – NA
■ Quality and Performance Management (7206ENG) – GC	■ Integrated Design Project (4003ENG) – NA, GC
	■ System Design Project (6209ENG) – NA
	■ Advanced Reinforced and Pre-stressed Concrete (7304ENG) – GC
	■ Project Management (7201ENG) – GC

NA = Nathan Campus; GC = Gold Coast Campus.

Each round of the survey involved a total of 32 closed (i.e., multiple choice) questions that related to students' lived experiences during the COVID-19 pandemic, including: (1) personal circumstances (nine questions); (2) learning experiences (eleven questions); (3) wellbeing responses (twelve questions). The "personal circumstance" section of the survey related to students' age group, gender, residential status (i.e., international, Australian permanent resident, Australian citizen), living arrangements, family situation, whether they had tested positive for COVID-19, household income, discipline and campus of study, and level and year of study. The "learning experiences" section of the survey related to the number of courses they studied in a particular trimester, experiences of final marks for their coursework in that trimester, overall average grade they obtained in that trimester as well as in the programme so far, access to technology during COVID-19, studying off-campus during COVID-19, online communication tools for coursework learning, and range of online teaching methods. The "wellbeing responses" section of the survey consisted of the standard Kessler Psychological Distress Scale (K10) test (Kessler et al., 2002), which is used internationally by professionals to yield a global measure of distress based on questions about anxiety and depressive symptoms that a person has experienced in the most recent four-week period.

A total of 85 and 35 students participated in the first and the second rounds of the survey, respectively. In the first round, 52 students completed the entire survey, whereas in the second round, 21 did. Therefore, the data from 52 and 21 students in the first and second rounds, respectively, were considered for the analysis. No incentives were offered for responding to the survey. The details of the survey were shared with the respondents. Participants completed a participation consent form before starting the surveys.

DATA PREPARATION AND ANALYSIS

An online LimeSurvey questionnaire was used to collect the data. Upon completion of data collection, the data was exported into.xlsx format. To conduct the analysis using IBM SPSS software, it was necessary to transform data from string to numerical format and then add them into an SPSS file. In the next step, data cleaning, integration, and labelling into proper groups were conducted, which were necessary prior to performing the statistical analysis.

The statistical analysis involved two main phases, namely, descriptive statistical analysis and inferential statistical analysis. The descriptive statistical analysis provided information about the immediate group of data and enabled presenting the data in a more meaningful way, which allowed improved interpretation of the data. Inferential statistical analysis allowed deeper analysis of the data and exploration of relationships between different variables within the sample (Taylor & Cihon, 2004). The tests that were used in inferential statistical analysis included the normality test, Chi-square test, Mann–Whitney, Kruskal–Wallis, and odds ratio test.

In the first trial of performing the inferential analysis on a normal set of data, no meaningful relationship was discovered due to the small sample size and large data dimensions. Consequently, in the second trial, the Kessler Psychological Distress Scale test levels and some question levels with more than four dimensions were reduced. However, no meaningful relationship was found. In the third trial, the rest of the question levels were reduced to two or three dimensions to enhance the potential for finding some relationships. This latest trial succeeded and some relationships among variables were found; these are presented in the following sections.

RESULTS AND DISCUSSIONS

The following sub-sections summarise key findings from analysing the first two rounds of survey data.

PERSONAL CIRCUMSTANCES

Table 10.2 summarises the first two rounds of analysis results with regard to the "personal circumstance" section.

As can be seen from Table 10.2, the majority of participants (i.e., more than 50%) were in the age group 21–30. This is followed by the age group of 'up to 20' (more than 20%) in Round 1. In Round 2 of the survey, the age groups of "up to 20" and "41 and above" showed the same percentage of participation (23.8%). More than 70% of the participants were male. Additionally, more than half of the participants (57.7% in Round 1 and 66.7% in Round 2) were Australian residents. The remaining percentage of students (42.3% in Round 1 and 33.3% in Round 2) were international students, who were on either student visas or other types of temporary visa.

TABLE 10.2

Summary of the Analysis Results with Regard to the "Personal Circumstance" Section of the Survey in the First and Second Rounds

Survey Characteristic	Possible Answers	Frequency (%)	
		1st Round	2nd Round
Age group	Up to 20	23.1	23.8
	21–30	63.5	52.4
	31–40	5.8	23.8
	41 and above	7.7	0.0
Gender	Female	26.9	23.8
	Male	73.1	71.4
Residential status	Australian resident	57.7	66.7
	International (study or other visa)	42.3	33.3
Family situation	I do not have a partner or close family member(s)	3.8	4.8
	I do not live with my partner/family, and they live in Australia	9.6	4.8
	I do not live with my partner/family, and they live overseas	28.8	19
	I live with my partner/family	57.7	71.4
Living arrangement	I live alone	15.4	14.3
	I live with one person	17.3	28.6
	I live with more than one person	67.3	57.1
Tested positive for COVID-19	No	96.2	90.5
	Yes	3.8	9.5
Household Income	No change	36.5	38.1
	It has gone down	48.1	47.6
	It has gone up	15.4	14.3
Discipline and Campus of study	Civil Engineering (Gold Coast)	59.6	33.3
	Civil Engineering (Nathan)	17.3	19.0
	Electrical/Electronic Engineering (Gold Coast)	7.7	0.0
	Electrical/Electronic Engineering (Nathan)	5.8	14.3
	Environmental Engineering (Nathan)	9.6	4.8
	Mechanical Engineering (Nathan)	0.0	4.8
	Mechanical Engineering (Gold Coast)	0.0	19
	Engineering Project Management (Gold Coast)	0.0	4.8
Level and year of study	Postgraduate – Mostly/all 1st year courses	9.6	9.5
	Postgraduate – Mostly/all 2nd year courses	7.7	4.8
	Undergraduate – Mostly/all 2nd year courses	40.4	33.3
	Undergraduate – Mostly/all 3rd year courses	32.7	42.9
	Undergraduate – Mostly/all 4th year courses	9.6	9.5

Whilst the majority of the participants (i.e., more than 50%) were living with a partner or family, a considerable percentage (i.e., 32.6% in round 1 and 23.8% in round 2) of participants did not have a partner or close family member(s) at all, or did not have a partner or family member(s) in Australia. More than 84% of students lived with at least one person, whilst a considerable percentage of students (i.e., 14.3% in round 1 and 15.4% in round 2) lived alone.

Whilst less than 5% of students had tested positive to COVID-19 or had a family member/housemate who tested positive to COVID-19 in Round 1 of the survey, the percentage had increased to approximately 10% in Round 2. More than 45% of students had reported that their household income had gone down during the pandemic. The majority of participants were from the Civil Engineering discipline at the Gold Coast and Nathan campuses and were in either 2nd or 3rd year of undergraduate study.

LEARNING EXPERIENCES

For this section of the survey, the study aimed to identify: (1) the main difficulties that participants have encountered during the pandemic; (2) their preferences of teaching methods and communication tools used for coursework learning; and (3) the overall impact of the pandemic on students' performance. The outcome of this part of the research is summarised in the following sub-sections.

Students' Main Difficulties during the Pandemic

In this study, the difficulties experienced during the pandemic were classified into two main groups, namely access to technology, and studying off-campus (Figure 10.1). Technology was specified as follows:

- Desktop/laptop.
- Smart (touch) screen (bigger than a phone).
- Webcam.
- Smart (digital) pen.
- Microphone/speaker.
- Scanner.
- Communication software (Teams, Blackboard, etc.).
- Internet download/upload speed.
- Poor internet access.
- Internet data limit.

As can be observed from Figure 10.1, the main difficulty that participants could not deal with is related to accessing a printer. For phases 1 and 2, 21.1% and 23.8% of participants, respectively, stated that accessing a printer was a huge problem that they could not deal with during the pandemic. This is followed by difficulty in accessing a smart screen that is bigger than a phone (i.e., 23.1% of participants in Phase 1 and 19% of participants in Phase 2) and accessing a scanner (i.e., 13.5% of participants in Phase 1 and 19% of participants in Phase 2).

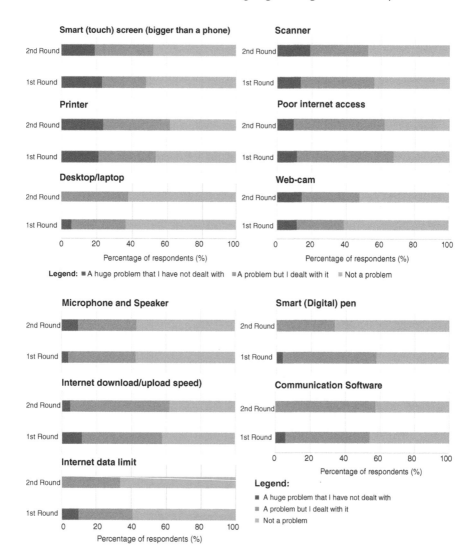

FIGURE 10.1   Students' difficulties in accessing to technology.

The main problem participants faced (regardless of whether they could deal with it or not) is related to poor internet access (i.e., 67.3% of participants in Phase 1 and 61.9% of participants in Phase 2). This is closely followed by internet download/upload speed (i.e., 57.6% of participants in Phase 1 and 61.9% of participants in Phase 2) and printer access (i.e., 53.8% of participants in Phase 1 and 61.9% of participants in Phase 2).

Figure 10.2 shows the outcome of the survey on students' difficulties in studying off-campus, where the difficulties were classified as follows:

- Noisy home environment.
- Elderly carer responsibilities.

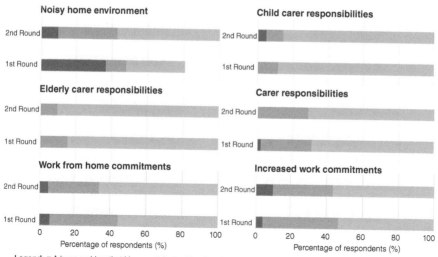

FIGURE 10.2   Students' difficulties due to studying off-campus.

- Work-from-home commitments.
- Childcare responsibilities.
- Sick partner/family member carer responsibilities.
- Increased work commitments.

As can be observed from Figure 10.2, the main difficulty that participants could not deal with relates to studying in a noisy home environment (36.5% of participants in Phase 1 and 9.5% of participants in Phase 2). This could be related to the fact that more than 84% of students live with at least one person in their household (see Table 10.2). The other main difficulties that participants could not deal with include increased work commitments and work-from-home commitments.

Furthermore, it can be seen from Figure 10.2 that the main problem participants experienced (regardless of whether they could deal with it or not) is related to a noisy home environment (i.e., 48% of participants in Phase 1 and 42.8% of participants in Phase 2). This is closely followed by increased work commitments (i.e., 46.1% of participants in Phase 1 and 42.8% of participants in Phase 2) and work-from-home commitments (i.e., 44.2% of participants in Phase 1 and 33.3% of participants in Phase 2).

Students' Preferences of Communication Tools and Teaching Methods

Figure 10.3 presents the outcome of the survey on students' preferences in using communication tools used for coursework learning, where the communication tools were considered to be the following:

- Blackboard – announcements.
- Blackboard – discussion.

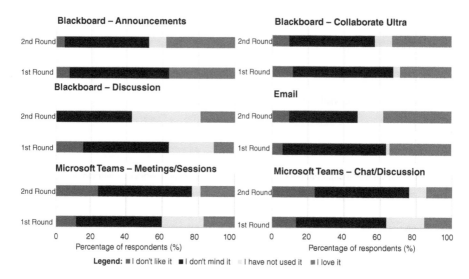

FIGURE 10.3 Students' preferences in using communication tools.

- Blackboard – Collaborate Ultra.
- Microsoft Teams – meetings/sessions.
- Microsoft Teams – chat/discussion.
- Email.

As can be observed from Figure 10.3, the main communication tool that students "love" is Blackboard – announcements (36.6% of participants in Round 1 and 38.1% of participants in Round 2). This is followed by email (34.6% of participants in Round 1 and 38.1% of participants in Round 2) and Blackboard – Collaborate Ultra (28.8% of participants in Round 1 and 33.3% of participants in Round 2). Furthermore, the results show that Blackboard – discussion was not used by a considerable percentage of participants (i.e., 25% of participants in round 1 and 38.1% of participants in round 2). Microsoft Teams (meetings/sessions) and Microsoft Teams (chat/discussion) are among the least preferred communication tools by the participants.

Figure 10.4 presents the outcome of the survey on students' preferences in the range of teaching methods.

In this study, the teaching methods were classified as follows:

- Live stream lectures/tutorials: This method enables students to join a live session, view the lecture as it is being held, interact with their peers and instructor, and participate in the class discussions.
- Pre-recorded lectures/tutorials: This is a pre-recorded lecture capture that is made available to students. This does not allow for a direct interaction with peers and instructor, nor for participation in the class discussions.
- Using the lecture time for discussion: In this method, the lecture materials and recordings are provided to students prior to the lecture session to study. The lecture time is then used to discuss the content of the lecture.

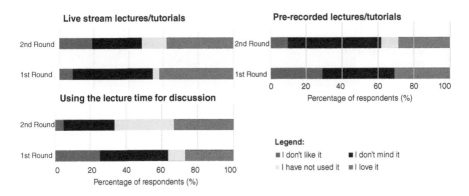

FIGURE 10.4   Students' preferences in range of teaching methods.

As can be seen in Figure 10.4, the participants' most preferred teaching method is live steam lectures/tutorials (42.3% of participants in round 1 and 38.1% of participants in round 2). This is followed by pre-recorded lectures/tutorials and using the lecture time for discussion.

Overall Impact of the Pandemic on Students' Performance

To investigate the impact of the pandemic on students' performance, three main aspects have been considered:

1. The number of courses students completed in comparison with the number of courses they originally planned to study in a particular trimester during the pandemic: This is to identify if the pandemic resulted in students completing fewer courses than they originally planned.
2. Students' perceived experience of their final coursework mark during pandemic in comparison with a previous grade given prior to the pandemic: This is to identify students' perceived impact of the pandemic on their performance.
3. Students' actual performance in a particular trimester (in terms of grade point average (GPA)) during the pandemic in comparison with their actual overall performance in the programme to date (in terms of Cumulative Grade point average (CGPA)): This is to identify the actual impact of the pandemic on students' performance.

Figure 10.5 shows the results of the survey on the number of courses that the participants have completed, compared with the number of courses they originally planned to study in a particular trimester during the pandemic. It can be observed from Figure 10.5 that 17.3% of students stopped studying in one or more courses in the first phase of the study; however, this number is reduced to 9.5% in the second phase of the study. This might be due to the fact that students have started to settle into the new learning environment and could plan better, given the experiences they had during the first trimester (i.e., first phase of this study).

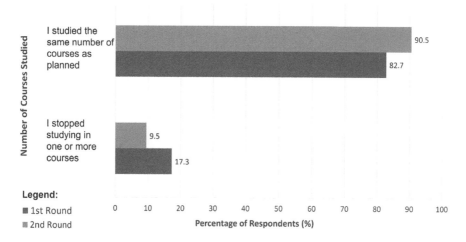

FIGURE 10.5   Number of courses completed versus number of courses originally planned to study.

Figure 10.6 presents the outcome of the study on participants' perceived experience of the final coursework marks in a particular trimester during the pandemic as opposed to those received in a pre-pandemic trimester. It can be seen from the figure that a significant percentage of participants (i.e., 57.7% of participants in Phase 1 and 66.7% of participants in Phase 2) believed that they did not do as well as they usually do. Furthermore, it is observed that 17.3% participants in Phase 1 and 9.5% of participants in Phase 2 of the study believed that they performed better than in the previous trimesters.

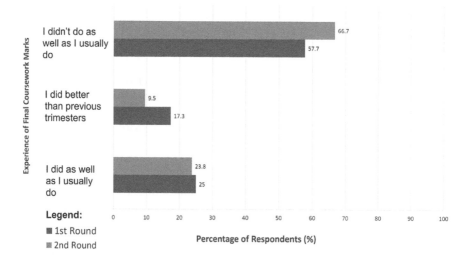

FIGURE 10.6   Students' perceived experience of final coursework mark during pandemic in comparison with the one prior to pandemic.

Figures 10.7a–b present the students' actual performance in a particular trimester (in terms of GPA) during the pandemic, and students' actual overall performance so far in the programme (in terms of CGPA), respectively. Comparing these two figures reveals that the number of students with a GPA of Fail (F) has increased during the pandemic. Whilst the percentage of students with a GPA of Pass (P) remains the same as that for the CGPA of Pass (P) in round 1, the percentage has significantly increased in the second round. Additionally, the percentage of students with the GPA of Distinction (D) has reduced, compared with the percentage of students with the CGPA of Distinction (D). The percentage of students with the GPA of Credit (C) and High Distinction (HD) has relatively remained the same as the percentage of students with the CGPA of Credit (C) and High Distinction (HD), respectively.

WELLBEING

COVID-19 has had significant implications for students around the world. Table 10.3 summaries the results of K10 tests and the survey on students' history of learning disorders and mental health issues. Of the students surveyed in this study, it is noted that approximately one in six students had a history of mental health issues (see Table 10.3). This prevalence is slightly lower than that of the Australian population in general (one in five) estimated by the Australian Bureau of Statistics (Australian Bureau of Statistics, 2020-21). Additionally, it is observed that approximately one in seven students (15.4%) and one in 20 students (4.8%) had a history of learning disorder in phase one and two of the study, respectively. This difference between the first and second phases might be due to the small number of participants in the second phase.

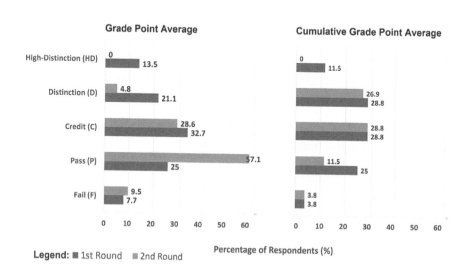

FIGURE 10.7   Students' actual performance in a particular trimester (in terms of GPA) during pandemic in comparison with their actual overall performance so far in the programme (in terms of CGPA).

TABLE 10.3

Summary of the Results of K10 Tests and the Survey on Students' History of Learning Disorders and Mental Health Issues

		Frequency (%)	
	Possible Answers	1st Round	2nd Round
K10 screening tool	Likely to be well	38.5	42.5
	Likely to be well or have mild mental health issues	50	52.4
	Likely to have moderate to severe mental health issues	50	47.6
History of learning disorders	Participants with a history of learning disorders	15.4	4.8
History of mental health issues	Participants with a history of mental health issues	17.3	19

What is striking is the portion of participants who are, at the time of the survey, likely to have moderate to severe mental health issues. The results of the survey presented in Table 10.3 show that one in every two participants is likely to have moderate to severe mental health issues. In other words, half of the participants warranted an in-depth mental health assessment at the time of the survey. It is notable that this percentage was similar during both rounds of the survey.

Chi-square test and Logistic Regression analyses were conducted to find the collection between moderate and high likelihood of mental health issues and a number of variables. Figure 10.8 presents the results of this analysis, which found a meaningful relationship between the moderate and high likelihood of mental health issues and a variable. Participants belonging to the following groups are significantly more likely to have possible mental health issues:

1. Age group of 21–30.
2. Not having performed as well as expected in some, most, or all courses.
3. No problem with microphone and speaker.

Conversely, the following variables had significant correlations with the group with low likelihood of mental health issues:

1. No problem with internet access.
2. No problem with internet data limit.

As mentioned previously, one of the limitations of this study is the small number of participants. Hence, interpretation of this data must be done with caution. Also, care needs to be taken when interpreting the results of correlations. Causational direction is not the primary result here. The geographic context of this study is another matter of importance. The experience of the pandemic in different parts of the world has been vastly different. Perhaps a more longitudinal study involving multiple centres and a larger population of participants can address some of these concerns in future.

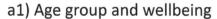

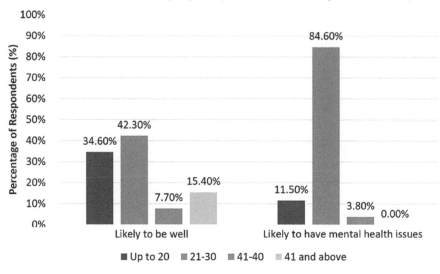

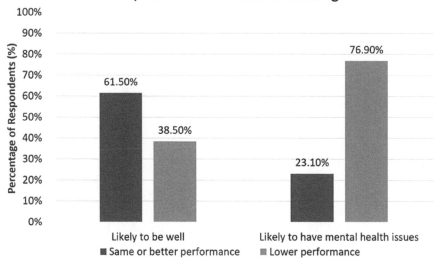

FIGURE 10.8 Correlation between students' wellbeing and some variables.

(*Continued*)

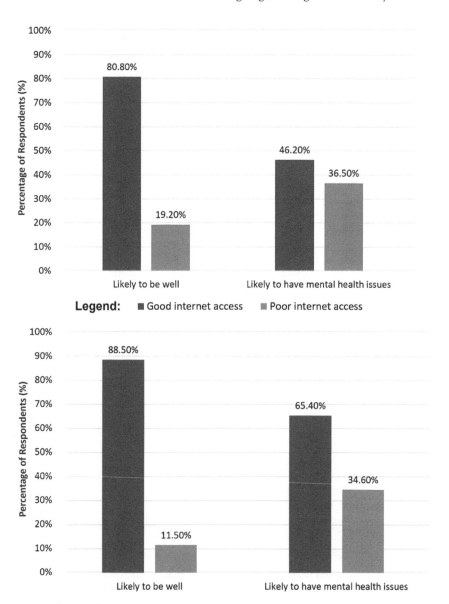

FIGURE 10.8 (Continued)   Correlation between students' wellbeing and some variables.

CONCLUSIONS

This study sought to understand the lived experiences of engineering learners grappling with the transition of curriculum delivery methods during the COVID-19 pandemic. A mixture of institutional, technical, and pedagogical challenges have been found that require different approaches to accessing content and communication to ensure sustained engagement and enhanced online educational experience. Through this study, it was evident that the learners faced myriad barriers relating to technology, off-campus study, and communication. The key technology-related difficulty that students experienced during the pandemic and could not deal with was related to accessing a printer. This was followed by the difficulty of accessing a smart screen that is bigger than a phone and the difficulty of accessing a scanner. The main off-campus study-related difficulty that students faced during the pandemic and could not deal with was related to studying in a noisy home environment. This was followed by difficulties resulting from increased work commitments and work-from-home commitments. The main communication tool that students preferred was Blackboard – announcements. This was followed by email and Blackboard – Collaborate Ultra.

The learners' accounts of their experiences of the sudden shift to largely online delivery methods suggest that their most preferred teaching method is live stream lectures/tutorials. While the pandemic resulted in an increase in the percentage of students with the GPA of Fail and a decrease in the percentage of students with the GPA of Distinction, nearly half or more of participants had a moderate to severe likelihood of wellbeing issues. Specifically, the age group of 21–30 was most likely to have experienced a high level of wellbeing issues. It is understood that students who reported to have lower performance compared to previous trimesters are more likely to have wellbeing issues. There was a meaningful relationship found between participants' age group, low performance (i.e., not performing as well as in previous trimesters), internet access and internet data limit, and the likelihood of having a wellbeing issue during the pandemic.

Analysis of the third round of survey will assist in drawing conclusions from this study. It is proposed that the findings will be immediately useful going forward in the COVID-19 pandemic, as ERT transitions to a new "business as usual" form of curriculum delivery. For example, educators from HEIs could use the findings relating to challenges for learners, preferred delivery, and communication methods to plan "next steps" accordingly for adjusting the student and educator experiences of online programme delivery.

REFERENCES

Abowitz, D. A., & Toole, T. M. (2010). Mixed method research: Fundamental issues of design, validity, and reliability in construction research. *Journal of Construction Engineering and Management, 136*(1), 108–116.

Alrefaie, Z., Hassanien, M., & Al-Hayani, A. (2020). Monitoring online learning during COVID-19 pandemic; suggested online learning portfolio (COVID-19 OLP). MedEdPublish, 9.

Australian Bureau of Statistics. (2020-21). *National Survey of Mental Health and Wellbeing.* https://www.abs.gov.au/statistics/health/mental-health/national-study-mental-health-and-wellbeing/latest-release

Bawane, J., & Spector, J. M. (2009). Prioritization of online instructor roles: Implications for competency-based teacher education programs. *Distance Education, 30*(3), 383–397.

Caldera, S., Desha, C., & Dawes, L. (2022). Applying Cynefin framework to explore the experiences of engineering educators undertaking 'emergency remote teaching'during the COVID-19 pandemic. *Australasian Journal of Engineering Education*, 1–13.

Creswell, J. W. (2013). *Research design: Qualitative, quantitative, and mixed methods approach.* Sage Publishing.

Dhawan, S. (2020). Online learning: A panacea in the time of COVID-19 crisis. *Journal of Educational Technology Systems, 49*(1), 5–22.

Henaku, E. A. (2020). COVID-19 online learning experience of college students: The case of Ghana. *International Journal of Multidisciplinary Sciences and Advanced Technology, 1*(2), 54–62.

Hill, K., & Fitzgerald, R. (2020). Student perspectives of the impact of COVID-19 on learning. *All Ireland Journal of Higher Education, 12*(2), 1–9.

Hodges, C., Moore, S., Lockee; B., Trust, T., & Bond, A. (2020). The difference between emergency remote teaching and online learning. *Educause Review, 27*, 1–12.

IBM Corp. (2016). *IBM SPSS Statistics for Windows, Version 24.0.* Armonk, NY: IBM Corp.

Kessler, R. C., Andrews, G., Colpe, L. J., Hiripi, E., Mroczek, D. K., Normand, S.-L., & Zaslavsky, A. M. (2002). Short screening scales to monitor population prevalences and trends in non-specific psychological distress. *Psychological Medicine, 32*(6), 959–976.

LimeSurvey. n.d. Retrieved from https://www.limesurvey.org/

Mahmood, S. (2021). Instructional strategies for online teaching in COVID-19 pandemic. *Human Behavior and Emerging Technologies, 3*(1), 199–203.

Mrug, S. (2010). *Encyclopedia of research design.* Thousand Oaks, CA: Sage.

Ponto, J. (2015). Understanding and evaluating survey research. *Journal of the Advanced Practitioner in Oncology, 6*(2), 168.

Rapanta, C., Botturi, L., Goodyear, P., Guàrdia, L., & Koole, M. (2020). Online university teaching during and after the COVID-19 crisis: Refocusing teacher presence and learning activity. *Postdigital Science and Education, 2*(3), 923–945.

Taylor-Guy, P., & Chase, A.-M. (2020). Universities need to train lecturers in online delivery, or they risk students dropping out. *The Conversation.* 1–5.

Taylor, J. K., & Cihon, C. (2004). *Statistical techniques for data analysis.* CRC Press.

TechnologyOne. (2020). *ANZ Student Survey Report 2020.* Retrieved from https://technologyonecorp.com/student-survey-2020

Xiong, W., Jiang, J., Mok, K. H., & Mok, H. (2020). Hong Kong university students' online learning experiences under the COVID-19 pandemic. *Higher Education Policy Institute – Blog.* https://www.hepi.ac.uk/2020/08/03/hong-kong-university-students-online-learning-experiences-under-the-COVID-19-pandemic

11 Reflections on Engineering Course Delivery

A Student Perspective

Lillian Guan, Hayden Ness, Shweta Mehta,
Andrew Busch, and Hugo G. Espinosa
Griffith University, QLD, Australia

CONTENTS

DOI: 10.1201/9781003263180-14

INTRODUCTION AND AUTHORS

The COVID-19 pandemic has proven to be a massive disruption to education and, although some places have returned to some semblance of normalcy, a lot of course content continues to be delivered virtually, with the possibility of this becoming standard. Students and educators alike have had to contend with frequent changes in social restrictions, and changes to course structure and mode of delivery. This added stress and uncertainty has been a challenging experience for all, and this is especially the case for students exposed to online and hybrid course delivery for the first time.

Since 2020, several studies have been conducted to examine students' experience of online delivery. These studies mostly revolve around self-report surveys of students and faculty staff. Some topics focused upon in these surveys were attitudes towards online learning; self-efficacy; preparedness for online learning, including technical and learning support; and assessment design (Asgari et al., 2021; Bogdandy et al., 2020; Ghasem & Ghannam, 2021; Jacques et al., 2020; Radu et al., 2020). Three particular studies accumulated observations across several courses (Ghasem & Ghannam, 2021; Jacques et al., 2020; Radu et al., 2020), while some drew upon sociological factors for distinguishing responses, such as gender and cultural background (Radu et al., 2020). In summary of these studies, students perceive that online learning enhances the flexibility and accessibility of courses, while also introducing setbacks such as difficulty in understanding instructions, technical issues, and motivational challenges. Over a variety of courses, those offering significant practical work resulted in greater disengagement from students and lower course satisfaction than more theoretical counterparts (Jacques et al., 2020). However, on average, all groups of the sample population scored higher grades than prior face-to-face offerings (Jacques et al., 2020). One reason for this may be the availability of new software for teaching, which students remarked was a positive outcome of learning online. Difficulty in focusing on study and the lack of a quiet study space were cited as the major disadvantages for students (Ghasem & Ghannam, 2021).

This chapter discusses the results of a survey delivered to Australian university engineering students who were likely impacted by COVID-19. The survey sought students' opinions, reflections, and recommendations with respect to improving course engagement while using online delivery. It is placed amongst the existing literature as a recent survey of current students and recent graduates, which aims to offer in-depth insight into students' needs. The survey responses offer student perspectives to educators, which are not always obvious, or fully understood. The discussion and analysis within this chapter aim to summarise key statistics, comments, and recommendations to assist educators to more effectively plan and develop courses that enhance the effectiveness of engineering education post-pandemic.

This chapter is interested in students' perceptions of course delivery. Although unconventional, we would like to first introduce ourselves as student authors of this chapter, to provide a background of who we are as you interpret our reflections about engineering course delivery:

- **Lillian Guan** – I am a sixth and final year student of electronics engineering and physics studies. Since my second year, I have been involved with university clubs and societies such as Women in Engineering and the Ladies in Technology, Engineering and Science Society at Griffith University. I also regularly volunteer as a mentor at Engineering Orientation and serve as a tutor and Peer-Assisted Study Session (PASS) leader, exposing myself to many interactions with students. In 2021, I completed my Honours thesis half-remotely due to COVID-19.
- **Shweta Mehta** – I am a recent engineering graduate who majored in software engineering. My university life was an experience I was very keen to share with soon-to-be and new university starters, and I was involved in multiple Griffith University initiatives (e.g., Uni-reach, STEM Squad, and Griffith University Orientation and Open Days) that would allow me to interact with just such a sub-group of students. I was President of the Women in Engineering leadership programme at Griffith University as well as Vice-President of Engineering for the Ladies in Technology Engineering & Science society. I have also been a lab demonstrator for a first-year engineering course for the last three years, which has kept me connected with students across engineering majors and different stages of the programme.
- **Hayden Ness** – I am a fifth-year Electronics and Computer Science student at Griffith University. Since enrolling in my degree, I have been very engaged with university life. I have always attended and engaged with my on-campus classes, assisted with various university initiatives, been executive and president of the engineering club, been a PASS leader for four years, and otherwise developed a good and active relationship with many on-campus engineering students and staff, in addition to students outside of engineering.

METHODS

SURVEY DISSEMINATION

The survey was distributed via two methods: first, an email containing a link to the survey was sent to all enrolled Griffith University engineering students and, second, links were shared on the authors' social media (e.g., LinkedIn). Students were informed that participation was voluntary, anonymous, and that responses would be used for the purposes of this research and deleted following the completion of the study. Approximately three weeks were given to complete the survey with reminders being sent periodically. A total of 65 completed responses were received.

Respondents were aged between 19 and 24 years old, predominantly male (75%), and enrolled in Electronic, Environmental, Civil, and Mechanical Engineering programmes. The majority of respondents were from Griffith University; of these, 61 were studying within Australia, three within China, and one within the UAE. Although the survey was not limited to Griffith University students, our capacity to

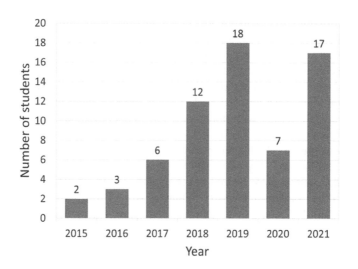

FIGURE 11.1   Year of enrolment of survey participants.

reach students enrolled in other academic institutions was limited. Respondents' initial year of enrolment in the Bachelor of Engineering spanned 2015 to 2021 (Figure 11.1). The proportion of respondents commencing in 2020 in the sample was quite low ($n = 7$). Although we do not specifically know why this is the case, it coincides with the lower rate of enrolment that occurred at the height of the COVID-19 pandemic, where the worst impacts on education were observed.

Survey Design

The online survey (see Appendix), created using Google Forms, aimed to assess engineering students' perceptions of, and suggestions about, the delivery of engineering courses in online and blended modes. The survey asked a mixture of quantitative and qualitative questions across the categories of learning activities (lectures, tutorials/workshops, laboratories), assessment and collaboration, and groupwork, which reflect the structural paradigm typical of most universities that may not change in the foreseeable future. A combination of multiple choice, five-point scale questions, and short response questions were asked, to gauge students' responses. The results analysed included qualitative evidence to support the quantitative data.

The survey was completely anonymous and took an estimated 15 minutes to complete. The research was approved by the Griffith University Ethics Committee under ethics approval code GU ref no: 2022/033.

RESULTS AND ANALYSIS

In terms of the respondents' mode of study (on-campus or online), and their ability to study on-campus, it was reported that 34 respondents studied on-campus and

31 respondents studied online. Figure 11.2 presents respondents' primary motivations for attending university. The majority attended university for the purpose of learning, while about half reported they attend just to get their degree, socialise and make friends, or seek opportunities. These are all significant and worth considering. Relatively few respondents selected any option exclusively; those that did, typically stated they attend university just to learn. The question was asked with the intent of revealing the relative importance of teaching quality, giving students opportunities and time to interact, and providing (or relating course content to) beneficial opportunities for motivated students.

In Figure 11.2, respondents were able to select more than one response; the percentages reported in this figure refer to the percentage of all respondents who selected each reason for attending university.

Figure 11.3 looks at the preferred learning styles of students using the VARK model (Fleming & Mills, 1992). Although the idea that there are different kinds of learners has been disputed by some researchers (e.g., Fleming & Mills, 1992), it is likely that learners will be more likely to remain engaged with course content when learning according to their preferred style. In any case, approximately three-quarters of the respondents indicated they were multimodal learners, with visual-kinaesthetic being the most popular combination. There is evidence for multimodal learning being the most effective (Firmansyah, 2021). It is also interesting to note that the relative ranking of the learning styles matches the relative ranking of the "Learning Retention Pyramid" (Fleming & Mills, 1992).

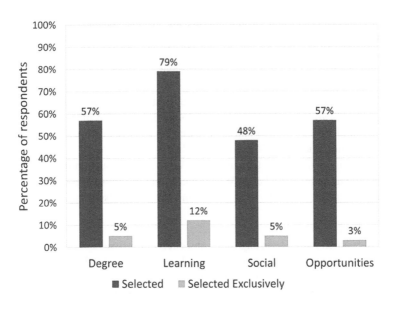

FIGURE 11.2 Reasons for attending university.

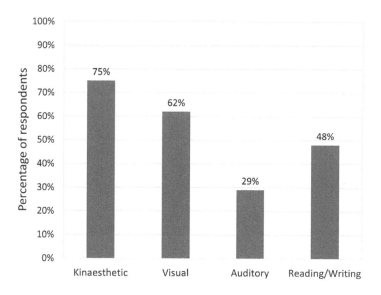

FIGURE 11.3   Preferred learning styles.

LEARNING ACTIVITIES

Lectures

Students were asked if they prefer single longer lectures or multiple shorter lectures for their courses. Eighteen students (28%) indicated a preference for longer, single lectures, 39 (60%) for multiple shorter lectures, and 8 (12%) had no preference. Figure 11.4 compares these results against the primary reasons for not attending, or not wanting to attend, live online lectures. The majority indicate a preference for multiple, shorter lectures, with the main reasons being problems with focusing, and preferring recordings. Of those preferring longer lectures, a relatively higher percentage cited having other priorities as a reason for their choice compared to those preferring shorter lectures. The responses suggest a strong preference for viewing recordings, with focus and commitment problems being the main driver. The survey also found that 25 (38%) students reviewed weekly content prior to lectures, 28 (43%) did not, and 12 (18%) sometimes did.

When prompted to answer the questions "What would make online lectures better and more engaging for you?" and "What other thoughts and suggestions do you have regarding lectures?" the following sentiments were expressed most frequently by respondents:

- 30% wanted interactive, engaging, and energetic delivery of content through the use of live demonstrations of problems, online whiteboards, in-lecture quizzes (including Kahoot[1]), or polls.
- 13% asked for good quality lecture recordings and technical finesse on the part of lecturers with the audio and visual quality in mind as well as the ability to skip sections and control the speed.

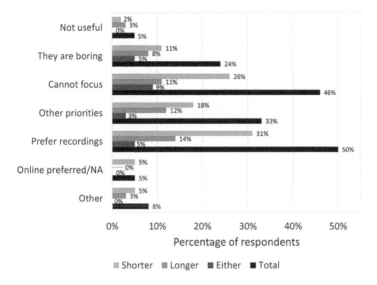

FIGURE 11.4   Preferred duration of lectures according to main reasons for not attending live lectures.

- 12% suggested shorter, well-paced lectures, even if there is more than one per week.
- 10% recommended lecturers turn their cameras on.

Other sentiments expressed by less than 10% of the participants included:

- A desire for more concise and summative lectures with an emphasis on engagement, the reasons for learning, additional information, and applications to the real world.
- A belief that online lectures offer improved work-life balance and avoid adversely impacting students who cannot attend in-person lectures.
- A preference for in-person lectures as it supports peer interaction, an important component of learning.
- A need for technological improvements: automatic captions, and recordings could have timestamps for improved navigation.

Tutorials and Workshops

Table 11.1 shows the results for the question asking students what they prefer to do in tutorials, from a selection of common or plausible activities, with the added option of providing a customised response. Multiple selections were permitted. An overwhelming majority indicated a preference for walking through practice problems, in addition to discussing lecture content and assessment. A minority asked for worksheets as a preference. Table 11.2 tabulates the students' main reasons for not attending online tutorials.

TABLE 11.1

Preference for Type of Activities during Tutorials

Activity	Count
Walk through practice problems	61 (94%)
Ask questions about content and assessment	47 (72%)
Discuss lecture concepts/content	45 (69%)
Worksheets alone	24 (37%)
Worksheets in groups	20 (31%)
Prepare for exams	1 (2%)

TABLE 11.2

Primary Reasons for Not Attending Online Tutorials

Reason	Count
I have other priorities	31 (48%)
I prefer to watch recordings	22 (34%)
I cannot focus	18 (27%)
They are not useful	13 (20%)
They are boring	10 (15%)
They are not assessed	9 (14%)

The following responses, in summarised and collated form, were given to the prompts "If you would like, you may clarify or explain your responses here" and "What other thoughts and suggestions do you have regarding the content, organisation, and delivery of tutorials?"

- Often, the help isn't needed, and so the tutorial shouldn't be assessed or compulsory.
- Online delivery prevents the ability to communicate problems with work effectively and receive immediate feedback or assistance.
- Due to the limited help available, some questions go unasked.
- Would like more Kahoot and engaging activities.
- Tutorial content should somewhat reflect and prepare you for exams, using similar types of questions.
- Weekly quizzes to help encourage studying and knowledge retention.
- Assessed tutorials add too much pressure to an already stressful workload.
- Pace can be too fast or shallow; explain in more detail.
- Tutorials should be recorded.
- Pre-recorded videos should be provided so that concepts can be revised as needed.
- Majority preference for in-person delivery; the immediate and specific help and interaction is important.

Laboratories

The responses indicated a strong preference for working in small groups: 38 students (58%) indicated a preference for working in small groups, 8 (12%) prefer working alone, and 19 (29%) did not mind either way. In addition, 40 students (62%) said that there is enough time in the labs to learn from and complete tasks, while 25 (38%) said there was not.

The following responses, in summarised and collated form, were given to the prompts "If you would like, please clarify or explain your previous responses here" and "What other thoughts and suggestions do you have regarding the content, organisation, and delivery of laboratories?"

- That laboratories should be in-person to develop students' hands-on skills or enhance learning.
- They do not often provide much learning and are just another hurdle.
- Labs are important or very beneficial.
- There is sufficient time to complete material.
- There is insufficient time to complete material or complete reports to a high standard.
- There is insufficient time or staff to receive help.
- I do not attend online labs due to lacking the software at home.
- I prefer working alone or in groups of two; a group of three is difficult to manage.
- I enjoy attending.
- Labs should begin with a slide outlining the lab tasks.
- Lab technicians are very helpful.
- Like online labs.

Authors' Comments

- Hayden – My preference is for one-hour-long, in-person lectures, but recordings are where I do most of my learning. I have difficulty even showing up to online lectures whereas I am always present at in-person lectures. For me, the difference is primarily social; I'm engaged when around my friends physically as it then feels like a team effort. I otherwise only use these in-person lectures as a primer, just getting the overall idea and contextual information, so that going through the recordings and other content to learn, and practice, is easier and more effective. While content coverage is most important, I'm usually most interested in why we are learning a topic, examples and stories of real-life applications, and other tangential elements; valuable information that is often omitted from the texts that I can read myself.
- Shweta – As a lab demonstrator myself I have seen first-hand the variance in attendance when an assessment piece is due and when it is not, as well as the benefit of recordings of live tutorials being provided to students. That is to say, if grades are at stake or there is a chance that something in the tutorial or lab will directly benefit an assessment piece, the students

will show up or find it online if it's available. For myself, I prefer to watch recordings of lectures unless I have friends that I can attend lectures with, particularly the lectures that contain solutions to labs or content directly related to other assessment. The real-life applications are a key for me too, which is why I find I can learn more in a lab than in a lecture; the experience of being able to play with the technology myself to explore and understand it exceeds that of just watching someone else do it when I can't follow along.

- Lillian – My preference is for fewer, longer lectures as it helps me to tie different ideas together and understand the connection between topics that I'm learning. For this reason, I attend my lectures in person, as it forces me to concentrate in the learning environment. While I prefer longer lectures, having modular supplementary course material such as lecture notes or video bites aids my learning too when I need the gist in a shorter timeframe. From my experiences, and observations of others, having the opportunity to work in groups or pairs during labs is ideal.

PLATFORM AND COURSE RESOURCES

It was found that 19 students (29%) prefer a weekly release of content, while 45 (69%) would like to have all resources at the beginning of the term, and 1 (2%) did not have any preference. This is particularly favourable amid an online learning environment that shows that most students take charge of their own learning.

Figure 11.5 looks at the most helpful types of content that educators can provide. There is an approximately even distribution between all types of content, except book readings. Practice questions and practice/previous exams were favoured the most.

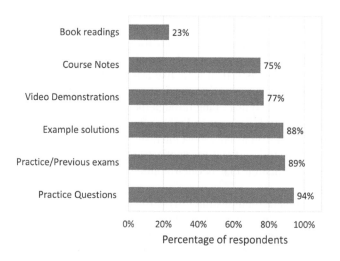

FIGURE 11.5   What type of resources are most valuable to you?

Comments concerning why book readings were least preferred are paraphrased below:

- The cost, time to obtain, or unavailability of a textbook makes it an unnecessary hurdle to learning that universities should be able to mitigate. The focus of students should be learning not jumping over hurdles.
- Book readings are too dense and time-intensive, which can be prohibitive in conjunction with the already large amount of time and energy that university requires. More focused and complete course resources are far more efficient for learning.

Authors' Comments

- Hayden – I don't have a particular preference for the form of resource, but I prefer having high-quality (clear, complete, correct, and focused) resources for each concept. I find this to be most effective for my learning as they are more easily digestible, help to track overall progress, and often more clearly delivered than in lectures. Generally, the less resistance I meet in the finding and understanding of resources, the more I engage with the course, and the more I will supplement my learning with external material. These days I can learn many topics easily, and freely or cheaply through books and online learning, and so feel that universities should at least offer value in the form of organised, clear, and time-effective delivery.
- Shweta – Despite having access to eBooks for a lot of courses, I have found them to be less useful than any of the other learning resources listed above. In fact, I've found that each of the learning resources listed can be very important at different stages in the course/assessment. I, like the peers I've discussed this with, believe that required reading should already be part of course content and taught, explained, and elaborated on by content deliverers. Additional individual learning aside, I wouldn't want to pay thousands of dollars to read a book by myself.
- Lillian – As the student sentiments reflected, I also rely heavily upon example problems and practice exams to reinforce my learning. Courses where I thrived contained a non-assessed practice component, such as weekly Wiley/NUMBAS quizzes, that allowed unlimited responses to build my confidence. While I use lectures as my primary tool, I enjoy having book readings to gain a comprehensive understanding of what I am learning. It is also a valuable opportunity to be taught from a different perspective.

Assessment

Figure 11.6 records respondents' ratings of perceptions towards group assignments ranked from 1 (*least favourable*) to 5 (*most favourable*). Overall, the results indicate that respondents perceive group assignments negatively.

This is an interesting statistic when compared to the sentiments students had about working in groups. Whether it concerns the size of the group, the accountability, or a mix of the two, students prefer to be able to work with the people they know but have

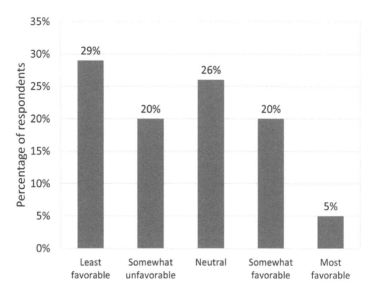

FIGURE 11.6   How do you feel about group assignments?

had strongly negative experiences with group assignments, which has definitely affected their attitudes towards future group assessments.

Authors' Comments

- Hayden – I am one of those students who has a strong dislike of group work. It has certainly aided me a couple of times, but I generally find it takes much longer to complete work, and at a reduced quality. I attribute this to the cost of communication between people, which is certainly amplified online. I also lose the ability to be as experimental and creative as I would like. I would accept more group work for larger projects where one person cannot reasonably complete it alone, there is a pre-established and fixed division of parts and responsibilities, and parts are individually demonstrable and marked. My preference is for formative assessments, which help me learn the content through practice and self-driven learning.
- Shweta – Unlike Hayden – and most of the respondents I suppose – I really enjoy group work. I find I can be more productive, motivated, and *motivating* when a group assignment is required. Of course, for myself it is because I naturally gravitate towards the delegation of tasks, enjoy the collaboration that comes with group work, and I know I have been lucky enough to pick my teams in most instances. However, I do sympathise with the majority's sentiment towards accountability in group projects – where I haven't been able to pick a team myself or have had to do an entire group's work myself the task has been daunting and infinitely harder to complete to the standard of a group's work.

- Lillian – Mirroring my fellow authors' comments, I enjoy group work at times when it is a technical problem that can be easily divided between members. Where creativity is concerned, I find it harder to negotiate the solution with other people's perspectives and, sadly, the process can be swayed where opinion/personality dominates the discussion rather than objectivity. I find group work exercises my ability to communicate with others and provide a balanced outcome, which is a soft skill favourably nurtured in students while at university, particularly in engineering.

Figure 11.7 records respondents' perceived ability to balance study time with assessments ranked from 1 (*most easy*) to 5 (*most difficult*). This question was asked as it was observed by the authors that students often spend all their time working on assessments but neglecting to study the remaining content that would better prepare them for the final exam or to achieve an overall higher course grade.

COLLABORATION AND GROUP WORK

Of the respondents, 41 (63%) stated that it is important for them to meet and study with their course peers, and 43 (66%) said it is important for them to meet their staff and lecturers in person. The results between the two questions are almost identical, meaning that almost everyone answered the same to both questions. Furthermore, these results indicate that approximately one-third of respondents did not rate in-person interactions as important for their course engagement. Some examples of in-person activities include studying alongside peers and having group discussions.

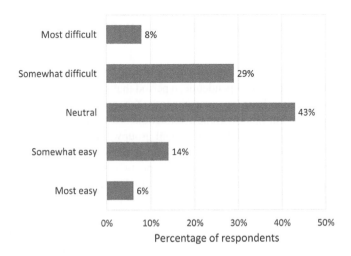

FIGURE 11.7 How hard do you feel it is to balance time spent studying with time spent on assessment?

While flexibility is conducive to online learning, most respondents suggested that they would prefer a blended learning environment, which allows them some face-to-face interaction, rather than being fully online. Respondents expanded that they would be willing to engage via common collaborative tools: discussion boards, shared notes, and online study groups. The demand for discussion boards correlates with requests for a discussion component in lectures and tutorials that is lacking in online delivery. It is unclear to us what "shared study notes" refers to, but it might suggest that respondents feel the provided learning resources are lacking in quality or quantity. Study groups are an effective way to learn and stay motivated, but this is often completely missing from online delivery.

OTHER

The final sentiments of the participants responding to the engineering survey call for more in-person learning activities and opportunities, and demand higher competence from the teaching staff. Studies, including the present one, suggest that effort to replace in-person practical work with engaging online alternatives is needed, as well as better accessibility to instructors or equivalent help, such as in the form of supplementary material or discussion boards (Jacques et al., 2020). In-person lab work and networking opportunities offer students reinforcement of the theoretical concepts and better engagement. Students also commented that lecturers need to be consistent in their delivery, whether it be using standardised tools such as Microsoft Teams or Blackboard[2] Collaborate Ultra, to avoid the technical issues that plague the online environment. More, too, is needed in the way of better-quality exams and assessment to counter the tendency to cheat and low motivation online.

However, students recognise overall the merit of flexible online delivery and appreciate the ability to take their education at their own pace. Final suggestions are provided in the conclusion section.

CONCLUSION AND SUGGESTIONS TO IMPROVE COURSE DELIVERY

This chapter gathered current engineering students' reflections on course delivery during the 2020–2021 COVID pandemic, a period that saw the development of either blended or fully online learning environments. Key areas that were investigated in this study were student demographics, reflections on content and assessment, and their perspectives regarding collaboration and groupwork. This work uniquely gathers anecdotal reflections of students' experiences during COVID and their recommendations for university teaching staff in a post-COVID scenario where blended learning is regarded with normalcy. A survey of 65 undergraduate engineering students was conducted within the framework of worldwide student evaluations of engineering course delivery, along with comparable results.

Overall, respondents reported a preference for the theoretical components of courses, traditionally covered in lectures, to be pre-recorded. In contrast, there is a preference for laboratories and tutorials to be conducted in-person. The main motivations for these activities include social wellbeing, increased engagement, increased opportunities for discussion, and the availability of help. Reflections highlighted that, since engineering is a hands-on degree, in-person engagement is essential.

Many of the qualitative responses also highlighted where improvements could be made to in-person delivery.

LECTURES

Students have indicated that there is a strong preference for watching recorded, rather than live, lectures, with the primary reasons being to assist with focus and avoid timetable clashes with other commitments. Consequently, several shorter weekly lectures seem to be preferred by the majority of students. For the present COVID scenario, this could be implemented by recording videos for the flipped-classroom approach and breaking down video demonstrations of foundational and longer concepts into segments; for example, a single video on how to do a cross product. By taking this approach, more time can be committed to interaction, discussions, and applied examples during lectures and tutorials rather than repeating basic concepts. This flipped-classroom approach is adopted widely amongst first-year mathematics and science courses to reinforce the basic concepts through easy access. By doing this, more time can be committed to interaction, discussions, and applied examples during lectures and tutorials, rather than repeating the foundational concepts. These concepts should still be covered as much as possible, but can be considered to be second in priority to students' questions.

Of course, producing these to a reasonable quality is a very difficult and time-consuming process. It should be sufficient in most cases to provide clips from previous lectures or external media that already provide a great coverage of a given concept. These few videos would be most effective if they are engaging and promote students' tendency to seek other learning material as needed. We believe that it is not important for the course creators to record an entire course like this by themselves in a single term.

Ideally, these few videos should be of high quality and presented in a manner consistent with the way in which the concepts are used throughout the remainder of the course. For instance, the usage of online whiteboards, simulation environments, or other technical software for problem solving would leverage content.

LABORATORIES AND TUTORIALS

Laboratories and tutorials provide a valuable opportunity for in-person interaction and practice, necessary to develop professional engineering skills. Encouraging laboratories to be completed in pairs but allowing individual work if the student prefers and the resources permit, would ensure that learning objectives are met while keeping students engaged. A finding of this study appears to be students' lack of engagement specifically with tutorials. Voluntary group work can also help to bridge gaps between students' understanding based on their progress through the lecture material, and laboratory instructors are subsequently afforded more time to deliver help while spending less time marking.

This leads into the frequent complaint that not enough help is available during laboratories and tutorials. The success of a laboratory or tutorial relies primarily on whether students leave the session feeling that they understood the content and were helped when they required it. This decreases in an online environment where communication is also much more difficult. It is, therefore, worth encouraging a culture of asking for help and empowering peer assistance so that some of the more basic issues students are having can be rectified. It was also mentioned that these labs should be allowed to be completed ahead of time so that more of the help in the session can be focused on those who need it most. The heavier reliance of students on in-person activities, therefore, requires sufficient staff training and learning support.

Although not specifically discussed in the survey, we recommend that giving students an additional week to write a report reflecting on learning in the laboratory or tutorial would be beneficial. This allows the lab time to be used purely for the practical application of content, giving students the necessary hands-on practice, leaving theoretical or reflective work to be completed outside of class time. This additional week would also allow more time for students to review notes, gain a better understanding of the content, and provide more insightful answers to questions, rather than writing just anything in order to complete the lab quickly and then forgetting about it.

While this recommendation contradicts those of most other work-from-home learning activities, tutorials and laboratories should continue to be delivered in-person, if possible, with tutorials remaining unassessed.

ASSESSMENT

Regarding assessment, students responded that practical projects, research projects, assessed labs, and weekly quizzes helped students the most, with mid-term exams and assessed tutorials ranking a close second. The following recommendations are made based on the survey responses, although we acknowledge that the types and amounts of suitable assessment will vary from course to course.

Practical projects with a self-learning and reflection component would be a highly effective assessment, and do provide the opportunity for hands-on experience, independent learning, and deeper understanding of the content. Practical projects would especially benefit kinaesthetic learners, who were found to be in the majority in the participant group for the survey. We think it is best that projects incorporate deliverables that are staged across the timespan of the project, so that the project does not compete with learning other content. Offering multiple project options would be ideal in catering to differing interests and minimising the capacity for students to copy one another.

Weekly quizzes are most suitable for content-heavy courses, and would sit within the collection of types of assessment for a course. While mid-term and final exams are formal summative assessment, weekly quizzes provide students the opportunity to reflect on their learning on a low-stakes basis and at low risk of failure, while providing regular checkpoints to gauge content retention. We think that low-stake weekly quizzes with higher-weighted quarterly or mid-term tests would make most students satisfied and more up-to-date with their learning. We recommend that frequent assessment pieces, such as weekly quizzes, should have lower weightings to encourage learning. In relation to final exams, a weighting between 30% and 50% was perceived to be reasonable, depending on whether the course emphasises theoretical knowledge or hands-on practical skills, with a higher weighting normally allocated for more theoretically focused courses.

AUTHORS' FINAL COMMENTS

- Hayden – While COVID was a very destructive force for education, it's also a great time for educators and students to come together and have those important reflections and discussions about the way education is delivered. I personally believe the biggest problems are systematic rather than with the delivery of any individual course, but improvements here can still have a large benefit for both students and staff. I haven't found the survey results to be particularly surprising given my own experiences and understanding of my peers, and I believe the recommendations made would be well received by the majority. No single decision will satisfy everyone, but I do think many courses are falling quite short of their maximum potential effectiveness and overall satisfaction from students and would do well to learn from active discussion with students and other educators.
- Shweta – While we have been able to represent only a sub-section of the student population, I believe that overall changes using the above recommendations would greatly benefit the general populace. The world of education adapted swiftly to the shift to online learning, and now the technology and methods used thus far can be refined. Pre-recording lectures, designing course-spanning projects, and setting up weekly quizzes is definitely more work, but it will result in a rise in content understanding and retention as well as student satisfaction. Of course, none of the suggestions we've received and recommendations we've made will fit *all* courses, but a dynamic and – dare I say – agile approach will greatly benefit an increasingly diverse student population.
- Lillian – In reviewing the current studies and our own experiences, it is evident that COVID has challenged both sides of engineering education. Educators now cater to a wider range of students' needs, and while students have more flexibility, they are disadvantaged by new learning barriers such as a lack of engagement with courses and lack of peer-to-peer learning that directly result from online learning. By implementing some of these recommendations, specifically adopting more in-person components where possible and varying assessment types and weights, I believe better preparation of engineering students as graduates in the profession is to come.

APPENDIX
SURVEY QUESTIONS

Demographic	
Which university do you attend?	**Griffith / UQ / QUT / TAFE / Other (specify)**
In what year did you begin your programme?	2015–2021
Do you usually study full-time or part-time?	Full-time / Part-time
Do you usually study online or on-campus?	Online / On-campus
What country are you studying in?	Qualitative response
Is attending campus possible for you?	Yes/No
What is (or was) most important to you at university? [multiple choice]	(Social/Friends) / Opportunities / Learning / (Degree/Jobs)
Which learning styles usually work best for you? [multiple choice]	Kinaesthetic, Reading & Writing, Visual, Verbal/ Auditory

Lectures	
Do you prefer a single long lecture, or multiple shorter lectures each week?	No preference / Multiple shorter / Single longer
Do you review weekly content prior to a lecture?	No / Yes / Other (specify)
What are your main reasons for not attending (or not wanting to) online lectures? [multiple choice]	They are not useful / They are boring / I cannot focus / (I have other priorities/Commitments) / I prefer to watch recordings / Other (specify)
What would make lectures better and more engaging for you?	Qualitative response
What other thoughts and suggestions do you have regarding lectures?	Qualitative response

Tutorials	
What do you prefer to do most in tutorials? [multiple choice]	Walk through practice problems / Worksheets in groups / Worksheets alone / Ask questions about content and assignments / (Discuss lecture content/ concepts)
If you would like, you may clarify or explain your previous responses here.	Qualitative response
What other thoughts and suggestions do you have regarding the content, organisation, and delivery of tutorials?	Qualitative response
What are your main reasons for not attending (or not wanting to attend) online tutorials? [multiple choice]	They are not useful / They are boring / I cannot focus / (I have other priorities/Commitments) / I prefer to watch recordings / They are not assessed / Other (specify)

Laboratories	
Do you usually find there's enough help in labs?	Yes / No
Do you usually feel there is enough time in labs to learn and complete tasks?	Yes / No
Do you prefer to do laboratories alone or in small groups?	Alone / Small groups / Either

(Continued)

Demographic

Which university do you attend?	**Griffith / UQ / QUT / TAFE / Other (specify)**
If you would like, please clarify or explain your previous responses here.	Qualitative response
What other thoughts and suggestions do you have regarding the content, organisation, and delivery of laboratories?	Qualitative response

Course Resources & Platform

Do you prefer resources to be made available each week, or all at once?	All at once / Each week / Other (specify)
What type of resources are the most valuable to you? [multiple choice]	(Course/Supplementary Notes) / (Previous / Practice exams) / Book Readings / Practice Questions / Video Demonstrations / Example solutions
Are you able to access required readings?	Yes / No / Only if freely available
If you would like, please elaborate on or clarify any of your responses here.	Qualitative response
What other comments and suggestions do you have regarding the type or availability of course resources?	Qualitative response

Assessment

How do you feel about group assignments for online courses?	1–5
How hard do you feel it is to balance time spent studying with time spent on assessments?	1–5
If you would like, please elaborate on or clarify any of your responses here.	Qualitative response
What other comments and suggestions do you have regarding the type of, and organisation of course assessment?	Qualitative response

Collaboration

Is it important to you to meet and study with your course peers?	Yes / No
Is it important to you to meet your lecturers/ staff in person?	Yes / No
What kind of collaborative tools/services would you be willing to engage with in your courses?	Qualitative response
What other comments and suggestions do you have regarding collaborative experiences and opportunities in your courses?	Qualitative response
Is there anything else you would like to say about how you think engineering courses should be organised and delivered online in the future?	Qualitative response

"Qualitative response" indicates open-ended questions.

NOTES

1 Kahoot is a free open-licence interactive platform that allows users to create quizzes. The use of Kahoot quizzes for education and training is growing in popularity in group contexts, e.g., lectures or tutorials to test students' knowledge. See kahoot.com.
2 Blackboard is a virtual learning environment and learning management system developed by Blackboard Inc. (https://www.blackboard.com/). It is currently the learning management system used at Griffith University, Australia.

REFERENCES

Asgari, S., Trajkovic, J., Rahmani, M., Zhang, W., Lo, R. C., & Sciortino, A. (2021). An observational study of engineering online education during the COVID pandemic. *PLoS One*, *16*(4), e0250041.

Bogdandy, B., Tamas, J., & Toth, Z. (2020). Digital transformation in education during COVID: A case study. In *2020 11th IEEE International Conference on Cognitive Infocommunications (CogInfoCom)*. IEEE, 173–178.

Firmansyah, B. (2021). The effectiveness of multimodal approaches in learning. *EDUTEC: Journal of Education And Technology*, *4*(3), 469–479.

Fleming, D., & Mills, C. (1992). Not another inventory, rather a catalyst for reflection. *To Improve the Academy*, 11, 137. Canterbury, New Zealand.

Ghasem, N., & Ghannam, M. (2021). Challenges, benefits & drawbacks of chemical engineering online teaching during COVID pandemic. *Education for Chemical Engineers*, *36*, 107–114.

Jacques, S., Ouahabi, A., & Lequeu, T. (2020). Remote knowledge acquisition and assessment during the COVID pandemic. *International Journal of Engineering Pedagogy*, *10*(6), 120–138.

Radu, M. C., Schnakovszky, C., Herghelegiu, E., Ciubotariu, V. A., & Cristea, I. (2020). The impact of the COVID pandemic on the quality of educational process: A student survey. *International Journal of Environmental Research and Public Health*, *17*(21), 7770.

Theme 4

*Insights on the Future of
Engineering Education*

12 Beyond COVID-19
Insights on the Future of Engineering Education

Maria Kovaleva
Curtin University, WA, Australia

Hugo G. Espinosa
Griffith University, QLD, Australia

CONTENTS

INTRODUCTION

During the two years since the COVID-19 pandemic began, humanity has learned to live with the virus. Lockdown measures and international travel restrictions have largely been lifted, although safety measures, such as social distancing, mask wearing, contact tracing, mandatory vaccination policies, and/or regular sanitising of shared facilities, remain in place in many countries. By early 2022, universities around the world were re-integrating elements of face-to-face (F2F) learning to re-establish academic vibrancy on their campuses and to promote the engagement and emotional wellbeing of students, faculty, and staff.

The COVID-19 pandemic necessitated extreme, unprecedented changes to learning that will have long-lasting consequences for higher education post-COVID-19. Although many universities were already gradually incorporating elements of online learning in their predominantly F2F delivery prior to COVID-19, the global pandemic dramatically accelerated this process, forcing unprepared educators to rapidly adopt new strategies and learn new technologies to cope with the change. Looking to the future, it is still unclear what the "new normal" for higher education will look like post-COVID-19, as institutions strive to provide contemporary, high-quality learning experiences that satisfy the expectations of students and their future employers, in increasingly austere economic times.

Brasco et al., (2019) discussed five major problems threatening global higher education:

1. Management of alternative learning modes.
2. Reduced government grants.
3. Concerns about the level of return-on-investment provided by higher education qualifications.
4. Reduced enrolment due to changing demography.
5. Consistent growth in services and operating costs.

These problems were identified prior to the onset of COVID-19, but are now particularly critical as we look to the future of higher education. In this chapter, we will present our projections for the future direction of education in a post-COVID-19 environment, specific to the context of engineering education. In particular, our discussion of the future of engineering education will highlight short- and long-term challenges for engineering education derived from the first problem identified by Brasco et al. (2019), pertaining to *management of alternative learning modes*. Despite this focus, our solutions to challenges derived from this particular problem will be proposed in consideration of the other four problems, given their interconnectivity. For each challenge we discuss, potential solutions and resources will be presented to assist engineering educators to meet these challenges. Towards the end of this chapter, we will focus on what we believe is currently one of the most important challenges for engineering educators: equipping engineering students to effectively support global efforts to achieve the Sustainable Development Goals (SDGs; https://sdgs.un.org/). Although the future of engineering education cannot be predicted with complete accuracy, projections for the future can be offered based on the influence of current and emerging technological, economic, and social developments (Froyd et al., 2014).

POST-COVID-19 CHALLENGES FACED BY UNIVERSITIES, EDUCATORS, AND STUDENTS

During the COVID-19 pandemic, remote learning and assessment facilitated the continuity of studies for engineering students throughout the world. For some educators and students, this relatively new and sudden shift brought positive changes, such as convenience of study, flexibility of participation, innovation in the engineering curriculum, and development of global universities. For educators who had less experience with online teaching, this abrupt transition increased their workload, stress, and burnout, and revealed incompatibilities with the traditional teaching system. From our perspective, a key challenge for engineering education moving forward is to embrace active and blended learning pedagogical strategies for online and hybrid delivery modes, while maintaining an environment of high-quality education. This can be achieved by incorporating the findings of evidence-based research, challenging the status quo of traditional engineering education, bringing industry practices into the classroom, and adopting advanced technologies for learning and assessment.

There is currently an abundance of information and communication technologies (ICT) available for educators to support online teaching, such as learning management systems and online engagement tools to facilitate video meeting, streaming, and content generation. The Top Tools for Learning 2021 website lists as many as 300 tools available for personal and workplace learning, and digital education (Hart, 2021). This massive expansion of supporting platforms reflects the overall level of demand for online learning ICT solutions, as well as the diverse needs of learners, educators, and programmes. In the case of online learning, it is clear that "one size does not fit all". We expect this list will continue to grow in the coming years.

Although individual circumstances vary, several challenges of remote learning due to COVID-19 are common for the majority of educators and learners around the world (Asgari et al., 2021; Qamar et al., 2019). We have collected 12 issues reported across the literature on digital engineering education, and have classified them according to five different categories:

1. Learning and engagement.
2. Assessment quality and academic integrity.
3. Teaching and pedagogy.
4. Research and industry.
5. Digital technologies.

Each category is discussed below, and a summary of the challenges is presented in Table 12.1.

TABLE 12.1

Identified Challenges of Digital Education and Proposed Solutions

#	Identified Challenges	Proposed Solutions	Reference
	i. Learning and Engagement		
1	Educator-learner interaction in a virtual classroom	Provide evaluative individual feedback; chat rooms and discussion forums for asynchronous discussions; AI-powered feedback	Boud and Molloy (2013) Brookfield (2006), Cavalcanti et al. (2021), Dow (2008), Grodotzki et al. (2021), Kochmar et al. (2020), and Seo et al. (2021)
2	The concept of university campus	Virbela; Second Life; Virtual and realistic 3D environments	Attallah (2020)
3	Reduced student engagement in collaborative learning, teamwork, and in communication with the institution	Active and problem-based learning; promote self-assessment; CDIO framework; flipped classroom	Clark et al. (2022), Cruz et al. (2020), Moore et al. (2017), Qadir and Al-Fuqaha (2020), and Qamar et al. (2019); www.cdio.org
4	Core professional qualities, such as interpersonal and practical skills	MOOCs and SPOCs; flexible curricula	Peimani and Kamalipour (2021) and Rouvrais et al. (2020)
5	Offering hands-on experience in a laboratory setting	Combine virtual and real laboratories, using VL in the first instance	IAOE/GOLC (2022) and Vergara et al. (2022)
	ii. Assessment Quality and Academic Integrity		
6	Contract cheating and other forms of academic misconduct in the absence of invigilation	IRIS; AI marking; reduce pressure of a single marked assignment by dividing it into sub-tasks	Alves et al. (2021), Tarigan et al. (2021), and Waltzer and Dahl (2022)
7	Verification of knowledge acquisition	Assessing project results by pre-recorded video/audio presentations	Grodotzki et al. (2021) and Khoo et al. (2018)
	iii. Teaching and Pedagogy		
8	Teaching staff training and support	Take an online class as a learner yourself; ask for feedback from experienced online instructors; LMS professional training for educators	Council of Australasian University Leaders in Learning and Teaching (2022)
	iv. Research and Industry		
9	Alumni engagement	Connecting on social media	Snijders et al. (2019)
10	Student retention	Summer schools, internships, and studentships	Bawa (2016) and Dias et al. (2021)

(Continued)

TABLE 12.1 (*Continued*)

Identified Challenges of Digital Education and Proposed Solutions

#	Identified Challenges	Proposed Solutions	Reference
		v. Digital Technologies	
11	ICT infrastructure	Offering flexibility Investment in teaching hardware	Hill and Lawton (2018)
12	Digital skills	Mandatory orientation programmes; CAD-based projects; remote desktop access and screen sharing tools	Attallah (2020) and Bawa (2016)

LEARNING AND ENGAGEMENT

Challenge #1 – Educator-Learner Interaction in a Virtual Classroom

One of the most obvious challenges when transitioning from F2F to online learning is the potential reduction of nonverbal communication during online interactions. For many people, the pandemic has emphasised the ways in which online communication limits the transfer of nonverbal information to which humans are accustomed, which occurs through tone of voice, gestures, facial expressions, and the synergy of group interaction. Removing these elements of communication is uncomfortable for human beings and, if not compensated by other methods, might reduce learning motivation, and increase feelings of loneliness, disconnect, and homesickness, especially for international students. Interactions between learners and educators are also significantly different in online classes compared to F2F classes. The live comments and questions posted by students in platforms such as iLecture[1] or Blackboard[2] Collaborate Ultra can be ambiguous or delayed, which causes confusion and distraction for the whole group. Perceptions about the quality of learner-educator interactions in online classrooms may also vary according to the cultural background of the participants. When learners and educators are from different cultural backgrounds, which is likely in online learning, values relating to hierarchical power distance and collectivism/individualism may cause variation in how learner-educator interactions are experienced (Damary et al., 2017; Tapanes et al., 2009).

These challenges can be addressed by incorporating a collection of techniques focussed on strengthening the connection between learners and educators. It was shown that it is highly effective for students when an educator provides evaluative individual feedback and uses chat rooms and discussion forums as a platform for conversations (Boud & Molloy, 2013; Brookfield, 2006; Grodotzki et al., 2021). These techniques can incorporate live interaction and asynchronous learning, since it is highly recommended that the live component should not be eliminated from online study (Dow, 2008). To accommodate large cohorts of students, delivery techniques can be designed that avoid increasing the workload of the educator (e.g., adding more teaching assistants to the teaching team). However, in the current climate within the education sector (Item 5 in the Introduction), it is unlikely that faculties will increase their quotas. Therefore, the more likely scenario is that artificial intelligence-based automated systems will be developed to provide individual student feedback (Cavalcanti et al., 2021; Kochmar et al., 2020; Seo et al., 2021). The research in this area began in the early 2000s and studies demonstrate that AI-generated customised feedback improves student learning outcomes and increases student performance (Johri, 2020).

Challenge #2 – The Concept of University Campus

Even though online learning is convenient and offers many advantages, it currently does not allow students, faculty, or staff to fully experience "campus life". University campuses provide a sense of community for many students; common shared spaces such as cafés, gardens, and libraries are places where students, academics, and general staff can interact. The campus environment is beneficial for students' wellbeing and is an important part of higher education that should not be replaced.

A few universities have already attempted to replicate campus life online, using immersive online environments such as Virbela (https://www.virbela.com/) and Second Life (https://secondlife.com/). The fast development of high-quality video cameras, audio technology, and head-mounted displays accelerates the acceptance of three-dimensional (3D) artificial worlds by institutions around the world. Virtual experiences of exhibitions, conferences, club activities, and virtual laboratories are now provided in a virtual world, where anyone around the world can "physically" join. It is a rapidly growing field that requires more financial investment in infrastructure and research on learning and mental health effects, as well as with issues relating to ethical behaviour and privacy.

Challenge #3 – Reduced Student Engagement in Collaborative Learning, Teamwork, and in Communication with the Institution

Left one-on-one with a computer screen, video recordings, and a textbook, students may experience feelings of stress, isolation, and insecurity (Hartman, 1995). From the educator's perspective, this can be perceived, at times, as an unwillingness to participate in online discussions and collaborative projects, which may dissuade the educator from further attempts to engage the student, exacerbating their initial feelings of isolation. This is a vicious cycle reported by students who transitioned to online study during the pandemic (Khan & Abid, 2021). This challenge also leads to low student retention rates, as described in further sections. The absence of hands-on experience and collaborative work on projects is especially incompatible with educating future engineers. The essential components of successful engineering education include the sharing of ideas with peers, demonstration of critical thinking, developing creativity, and learning communication skills.

Multiple studies show that active and problem-based learning approaches are excellent methods for achieving deep learning, improving practical experience, and developing complex problem-solving skills, which help to increase graduate employability (Qamar et al., 2019). In addition, students need opportunities to become competent in self-assessment (Brookfield, 2006; Cruz et al., 2020; Moore et al., 2017; Qadir & Al-Fuqaha, 2020), a reflective practice that is a necessary skill for engineers. A framework that has been successfully implemented in engineering courses worldwide is referred to as Conceive-Design-Implement-Operate (CDIO; http://www.cdio.org/; Qamar et al., 2019). Previously introduced at later stages of an undergraduate degree, it can also be added to first-year units to make use of interactive technology. Flipped classroom is another highly successful teaching approach that is compatible with online learning (Clark et al., 2022). Using these techniques in combination with advanced hardware and software technologies can assist in overcoming the issue of disengagement in remote learning.

Challenge #4 – Core Professional Qualities, such as Interpersonal and Practical Skills

Core professional competencies, such as interpersonal and practical skills, are developed via social learning processes, for instance, observing peers and faculty members. The value of leadership and interpersonal skills, which enable people to successfully collaborate and communicate with one another, are increasingly desired by employers and engineering bodies. Personal characteristics that build resilience to stress and contribute positively to academic performance, such as psychological capital, are also primarily learnt via social learning processes. Developing these interpersonal skills in online environments is possible but requires adjustments to teaching strategies (Biggs & Espinosa, 2021).

Valuable practical skills are acquired when students become familiar with requirements to follow safety lab practices and engaging with industry-standard measurement setups, even if they are not being used first-hand, which happens during group laboratory activities. The value of being situated in the environment of an engineering school should not be underestimated and must be considered by the teaching staff when redesigning curriculum for the future. Online education can easily lack social and vicarious learning opportunities.

Engineers will require new professional qualities in the post-COVID-19 world, which appears to be operating under high levels of volatility, uncertainty, complexity, and ambiguity (Rouvrais et al., 2020). These new professional qualities are ICT communication skills, adaptation to change, and empathy (Peimani & Kamalipour, 2021; Rouvrais et al., 2020). Emerging commercialisation programmes for postgraduate students and industry-oriented government grants also emphasise the importance of entrepreneurial skills for undergraduate engineering students.

The combination of massive open online courses (MOOCs) and small private online courses (SPOCs) might be the way in which "V-shape" skills will be trained in future engineers (Rouvrais et al., 2020). As opposed to the "T-shape" skills, which represent a good amount of general knowledge and one area of specialisation, "V-shape" skills represent spiral learning, where the expertise covers multiple disciplines and depth at the same time, resulting in well-rounded engineers with balanced knowledge and skills. Fundamental broad skills, which are usually taught in the first year of an engineering degree, may be provided with less teaching support and more time flexibility for their completion, such as through MOOCs. Once these skills are acquired, small-group, domain-specific courses that have extensive laboratory component and live interaction with the educators can be undertaken. These can be SPOCs, in-person, or hybrid classes. Industry will play a pivotal role in determining whether this system will dominate in engineering education, given that the final aim of higher engineering education is to increase graduate employability and address professional needs. To improve the chances of industry acceptance of this engineering education mode, it would be reasonable to actively engage industry and research companies in engineering curricula redesign.

Challenge #5 – Offering Hands-on Experience in a Laboratory Setting

Offering hands-on experience to engineering students is imperative; the more hands-on experience students are exposed to, the better prepared they will be for industry. During the COVID-19 pandemic, students either partially or completely lost the opportunity to interact with real equipment, due to the initial unavailability of suitable laboratory experiments for remote teaching. Espinosa et al., (2021) and Iyer et al., (2021) offered some alternative options for setting up remote labs and providing students with hands-on online practice. Other promising strategies for providing remote laboratory experience include the introduction of virtual reality and augmented reality laboratories, also referred to as virtual laboratories (VL) or virtual learning environments (VLE; Vergara et al., 2022). The opportunity for individual practice with the equipment propels the acquisition of knowledge and provides a multitude of advantages: reduced stress from interacting with high-cost equipment, low investment of time and budget, and no restrictions on time when interacting with the facility. However, VLs should not completely replace hands-on experiments; research has shown that a very high percentage of respondents prefer real laboratories or a combination of real and virtual laboratories (Vergara et al., 2022). Innovation in VL environments can be encouraged via the support of small grants and awards; for example, Global Online Laboratory Consortium (GOLC) annually awards the best online laboratories in three categories: visualised experiments, simulated experiments, and remote-controlled experiments (IAOE/GOLC, 2022).

ASSESSMENT QUALITY AND ACADEMIC INTEGRITY

Challenge #6 – Contract Cheating and Other Forms of Academic Misconduct in the Absence of Invigilation

Online assessment design is another major challenge that university educators have faced since the pandemic began. In particular, designing high-quality online assessment that is not highly susceptible to academic misconduct remains a significant concern for universities, despite the development of strategies to enhance academic integrity and technological resources to reduce cheating (Krambia Kapardis & Spanoudis, 2022).

During COVID-19 lockdowns, traditional F2F examinations (i.e., invigilated and time-limited) were largely replaced by virtual examinations, a practice that universities are likely to continue to implement in the future (Fluck, 2019). The lack of invigilation in online exams allows for the possibility that students may visit websites such as Chegg, ClutchPrep, CourseHero, Koofers, Quizlet, and SparkNotes for consultation during the exam. These websites provide students with opportunities to seek help from other people (online tutoring), as well as attain the answers to sample exam questions and textbook problems; research suggests their use is growing considerably (Nguyen et al., 2020). Chegg seems to be one of the most common online resources, as it allows real-time support from tutors, which assists students' study but is problematic when students misuse these services to cheat on exams designed to assess their level of knowledge (Raje & Stitzel, 2020). To counteract this issue, universities have invested in e-proctoring resources, such as ProctorU (https://www.proctoru.com/) and Intelligent Remote Invigilation System (https://www.irisinvigilation.com/). These are online proctoring services that allow students to complete their assessments online, using a camera to detect specific body movements indicative of cheating intentions. However, these proctoring platforms have been scrutinised due to privacy concerns, data breaches, and false cheating allegations (Henry & Oliver, 2022). Other simpler alternatives include monitoring timestamps from the students' access logs (Bilen & Matros, 2021). While these solutions can be effective to some extent, they cannot completely prevent students from cheating, and the probability of students being caught with significant evidence to indicate cheating is minimal (Bilen & Matros, 2021). These issues raise important questions about how universities can provide high-quality assessment that not only achieves desired learning outcomes but also effectively minimises cheating and academic misconduct.

Although potential solutions are still being debated, there are some evidence-based strategies educators can undertake to improve academic integrity:

- Asking students to sign an academic integrity declaration before the examination, where they have to confirm that the work being completed is their own (Krambia Kapardis & Spanoudis, 2022). Instructors can remind students of academic integrity policies during the lectures and before exams.
- Random drawing of questions from large databases, and changing the sequence of questions (de Sande, 2015). One available option is the platform NUMBAS, a free and open-source e-assessment system (https://www.numbas.org.uk/).
- Prohibit backtracking. This allows students to solve one questions at a time. According to Raje and Stitzel (2020), this strategy must be used with care to avoid adversely impacting the performance of students who have disabilities and may require access to the entire exam.
- Grading step-by-step solutions in mathematical problems. Students seem to be less likely to cheat if they know their working solution will be marked rather than just the numerical result (Janke et al., 2021).
- Converting questions into images and adding watermarks. These strategies impede students' capacity to copy and paste questions into web search engines, and uploading screenshots to web tutoring services, as these are flagged by the watermark (Raje & Stitzel, 2020).
- Delay score availability. When exams are available for large periods of time (i.e., 24 or 48 hrs), students who have finished the exam can provide correct answers to those who have not yet completed the exam (Smith, 2020).

In addition to the above strategies, it is important to design assessment pieces and create learning environments that promote academic integrity. This can be achieved by Setting clear expectations regarding academic integrity, having authentic, supportive educator-learner interactions and developing formative assessments that scaffold students' learning (Gottardello & Karabag, 2022). As an example, including multiple, short weekly assessments instead of one or two highly weighted assessments can minimise the likelihood that students will cheat in response to performance pressure (Alves et al., 2021). Encouraging students to develop learning goals also minimises cheating, as students consider it a "costly shortcut that undermines true understanding" (Janke et al., 2021, p. 7).

Challenge #7 – Verification of Knowledge Acquisition

In engineering courses, there are a few common assessment types, such as laboratory reports, multiple-choice tests, and written invigilated exams. Irrespective of the type of assessment employed, accurately verifying whether anticipated learning outcomes have been achieved is a difficult task, even in the absence of academic misconduct. Verifying knowledge acquisition has been especially challenging due to the transition to online learning and inability to conduct in-person assessment. This problem can be resolved by redesigning assessment types. For example, Grodotzki et al. (2021) discussed the use of self-assessment activities (e.g., weekly quizzes) as a means for testing students' knowledge acquisition, which also increases students' autonomy and intrinsic motivation. Importantly, motivation to complete the self-assessment was highest when they were voluntary and there was an incentive for completing them (e.g., bonus marks towards a student's final grade).

Verification of knowledge acquisition can also be enhanced through assessment-related activities designed to engage students in active learning and problem-based learning (Grodotzki et al., 2021). Active learning has been successfully promoted by flipped classroom (Khoo et al., 2018). Assessment items drawing on principles of problem-based learning can successfully assess knowledge acquisition when: (a) the project difficulty is aligned with the level of study, and (b) the task is novel enough that the solution cannot be easily found in Internet searches. Assessment drawing on principles of active and problem-based learning need not be conducted in-person; rather, it can be conducted virtually either synchronously or asynchronously by way of pre-recorded videos. For large cohorts of students, when marking individual presentations is infeasible, the assessment can be conducted for groups.

TEACHING AND PEDAGOGY

Challenge #8 – Teaching Staff Training and Support

Every teacher is unique; therefore, no single teaching technique will uniformly work for everyone. Although the transition online during the pandemic was smooth for some teachers who had previous online experience, it was challenging for others with little to no online experience. Online and hybrid education seems to be vital for the future of engineering education; with sufficient training and support, every educator around the world will be able to translate their teaching strategies and methodologies to the digital audience.

There are several ways in which educators can support themselves in transitioning to technology-based teaching. One effective technique for educators is to take an online class as a learner and reflect on which teaching methods worked and which did not work. Another powerful method is to ask for feedback on your online unit from experienced online instructors. Finally, it is important to remember that the principles of learning design remain the same for online, F2F, and hybrid mode:

1. Understand the level and diversity of the students.
2. Create learning activities that lead to attaining the learning outcomes.
3. Provide necessary means for the students to complete the learning activities.
4. Practice constructive alignment (Council of Australasian University Leaders in Learning and Teaching, 2022).

It is imperative that teaching staff are supported by their home universities and local teaching community. Professional training for educators in the advanced use of learning management systems (LMS), such as Blackboard, Articulate, and Canvas, can be very helpful. On a discipline-specific level, the international community shares their findings through societies and associations, as well as in peer-reviewed conferences and journal publications. Some of these include:

Societies and Associations:

- Australasian Association for Engineering Education (AAEE).
- IEEE Education Society.
- American Society for Engineering Education (ASEE).
- Research in Engineering Education Network (REEN).
- European Society for Engineering Education (SEFI).
- International Technology and Engineering Educators Association (ITEEA).
- African Engineering Education Association (AEEA).
- The Asian Society for Engineering Education (AsiaSEE).
- International Centre for Engineering Education (ICEE).
- Institute for the Future of Education (IFE)
- International Association of Online Engineering (IAOE).
- The Global Online Laboratory Consortium (GOLC).

Peer-reviewed journals:

- Journal of Engineering Education (Publisher: Wiley).
- Australasian Journal of Engineering Education (Publisher: Taylor & Francis).
- European Journal of Engineering Education (Publisher: Taylor & Francis).
- Transactions on Education (Publisher: IEEE).
- Transactions on Learning Technologies (Publisher: IEEE).
- IEEE Access: IEEE Education Society Section (Publisher: IEEE).
- Ibero-American Journal of Learning Technologies (Publisher: IEEE).
- Teaching in Higher Education (Publisher: Taylor & Francis).
- International Journal of Mechanical Engineering Education (Publisher: Sage).
- International Journal of Electrical Engineering Education (Publisher: Sage).

- Innovations in Education and Teaching International (Publisher: Taylor & Francis).
- International Journal of Electrical Engineering Education (Publisher: Wiley).
- Research in Engineering Design (Publisher: Springer).
- Journal of Civil Engineering Education (Publisher: ASCE).

International conferences:

- IEEE International Conference on Teaching, Assessment, and Learning for Engineering (TALE).
- American Society for Engineering Education (ASEE) Conference.
- Australasian Association for Engineering Education (AAEE) Conference.
- Frontiers in Education (FIE) Conference.
- Global Engineering Education Conference (EDUCON).
- World Engineering Education Conference (EDUNINE).
- Learning with MOOCs.
- International Symposium on Accreditation of Engineering and Computing Education (ICACIT).
- Technologies Applied to Electronics Teaching Conference (TAEE).
- Annual Conference of the European Society for Engineering Education (SEFI).
- International Conference on Interactive Collaborative Learning (ICL).

RESEARCH AND INDUSTRY

Challenge #9 – Alumni Engagement

Graduates of traditional institutions have a strong connection with their alma mater. They support their universities through financial sponsorships, guest presentations, panels, career fairs, donations, mentorships, and by being ambassadors for their universities. Alumni engagement with their universities can provide views on the relevance of programme structures and course content; it can also contribute with job opportunities for students (Ebert et al., 2015). Universities maintain their engagement by giving them designated places in their websites, sharing their stories on social media platforms, celebrating graduate successes through alumni awards, and promoting them as role models for current students. It was found that an important driver of alumni engagement was the students' level of trust in the quality of their relationship with staff (Snijders et al., 2019).

However, there is a concern that an online education model may reduce alumni loyalty and giving behaviour. Early alumni engagement through social media can improve this situation. Ding and Riccucci (2020) state that alumni have a strong sense of honour, making them responsive to the needs of their alma maters. As an example of this, alumni from Wuhan University in China provided faculty and students with supplies of food and water while they were locked in their residences at the early stages of the pandemic (Ding & Riccucci, 2020). Alumni engagement is an important avenue for further research post-COVID-19, as little is known about the long-term impact of online studying on alumni engagement.

Challenge #10 – Student Retention

Another big challenge associated with online education is student retention. As discussed in Bawa (2016), there are many reasons for attrition amongst online students. These include frustration with technology, the loss of focus and confidence due to the absence of formative feedback prior to the assessment, and lack of administrative and financial support (Dias et al., 2021). More specific to engineering, according to Geisinger and Raman (2013), attrition rates can be related to inadequate high-school preparation (i.e., maths and physics background); lack of engagement with course content and course delivery; lack of self-efficacy and self-confidence; and poor academic achievement. These issues are more visible in first-year programmes, which are common to all engineering disciplines and where students may have conceptual difficulties with core courses.

Theory and practice are strongly linked in engineering education, and the sudden transition to online teaching due to COVID-19 impacted the delivery of F2F hands-on activities. In order to adapt to the situation, whilst still satisfying curriculum requirements and addressing student needs, educators around the world employed several strategies, including remote integrated platforms, mobile learning, video experiment recording, and laboratory simulation (Darby & Lang, 2019; Espinosa et al., 2021). However, while some students adapted well to the online transition, others struggled and felt both disengaged and demotivated, causing losses in student retention (Means et al., 2020; Park et al., 2020; Ramo et al., 2021). The easing of restrictions at the end of 2020 allowed several institutions to adopt blended learning strategies, where educators combined synchronous and asynchronous modes of delivery, allowing students to undertake courses online and in-person (Espinosa et al., 2021; Rahman & Ilic, 2018).

The implementation of well-known pedagogical strategies, such as flipped classroom (Furse & Ziegenfuss, 2020; Schwerin et al., 2021; Talbert, 2017) and project-based learning (Gratchev et al., 2018; Kokotsaki et al., 2016), has increased in online courses since the start of the pandemic. According to Ramo et al. (2021), flipped classroom benefits from combining both synchronous and asynchronous delivery modes. If applied correctly, both techniques can have a positive impact on student retention. Vesikivi et al. (2020) describe how a project-based curriculum significantly improved student retention, particularly in first-year programmes, creating opportunities for teamwork and teacher-student interaction. Ryan and Reid (2016) reported a year-long controlled study of flipped classroom in a second-term general chemistry course and showed how its implementation led to improved student performance, satisfaction, and retention. When implementing flipped classroom, instructors should consider breaking up content into small pieces, delivering live sessions where students can ask questions and interact with the instructor and other students, and using real-world examples to illustrate course content (Ramo et al., 2021). Of course, for this technique to work, students need to be engaged and complete the required work prior to class. One path to achieve student engagement is through intrinsic and extrinsic motivation as described by Kosslyn (2021).

A review of 45 peer-reviewed articles published in a period of 10 years was conducted by Kuley et al. (2015), who reported potential causes of high attrition rates in engineering universities, as well as strategies to improve students' learning experiences. According to their findings, factors associated with engineering student attrition occurs at three levels: college-level, instructor-level, and student-level. Specific factors included first-year engineering curricula, mentorship, learning style, teaching delivery, peer influence and sense of belonging, motivation, academic achievement, gender, race, and faculty-student interaction. Considering these factors is critical for improving student retention rates.

DIGITAL TECHNOLOGIES

Challenge #11 – ICT Infrastructure

Online delivery and attendance cannot be possible without the essential technology for remote learning. This implies access to the Internet and having a mobile device, a laptop, a tablet or a desktop with a web camera and a headset with a microphone. Software packages are also necessary and may include specialised programming and CAD tools. Access to hardware and software is a major challenge, as educators and students who lack such resources are penalised, further deepening the "digital divide" (Hill & Lawton, 2018). Also, in an ideal situation, it is assumed that a learner has a quiet space to study, which might not always be the case. This potentially disadvantages students from vulnerable and underprivileged backgrounds. It would be a responsibility of the university to instruct students about the issue and provide solutions (e.g., common study areas and ICT).

Technology fails; therefore, online education should provide as much flexibility as possible regarding attendance and assessment submission to accommodate unexpected technology malfunction. For educators, universities must budget for investment in professional quality audio and visual (AV) infrastructure, as well as the AV engineers necessary to support the development and maintenance of high-quality educational resources.

Challenge #12 – Digital Skills

Digital natives, the generation that have grown up with ICT, and digital immigrants, those who acquired digital skills later in life, interact with technology differently. This gap is especially noticeable in the education sector and, if overlooked, may seriously disadvantage certain students. Poor digital skills of students and instructors not only make for a poor educational experience but also lead to a higher rate of student failure and even degree dropout. Since digital skills can be easily overestimated, both students and instructors should complete mandatory orientation programmes to improve their technological communication skills (Bawa, 2016). This can be offered as a marked component of the study to encourage participation, and for instructors, should be an essential part of their promotion procedure.

Finally, acquiring computing skills is an essential component of modern engineering degrees. In F2F mode, students attend computer laboratories and work on provided activities using specialised software with the guidance and immediate support of laboratory instructors. These activities can include programming, software design, computer-aided design (CAD), and data science (Attallah, 2020). Transferring this important teaching component to an online mode is challenging and can be accomplished if an instructor has remote access to the student's computer, which can be done by Team Viewer or via screen sharing. To further assist the students, step-by-step written instructions can be provided along with live sessions where an instructor performs the same steps, sharing the screen with the students. These sessions can be recorded for future use in asynchronous learning.

ACTIVE LEARNING AND SUSTAINABLE DEVELOPMENT GOALS (SDGS)

In 2015, the United Nations General Assembly released the Sustainable Development Goals (SDGs) to be achieved by 2030 (https://sdgs.un.org/). The 17 goals aim to address the most important and complex challenges the world is facing, such as poverty, inequality, clean water, climate action, and quality education (see Figure 12.1). Engineering plays a very important role in ensuring that many of these goals can be achieved. Goals associated with industry, innovation, and infrastructure; clean water; climate action; sustainable cities; and clean energy can be addressed through engineering solutions, by the engineering workforce. Universities play a critical role in ensuring the next generation of engineers is well prepared to meet the demands and challenges imposed by the SDGs.

The Australia, New Zealand, and Pacific Network of the Sustainable Development Solutions Network (SDSN; https://ap-unsdsn.org/) supports and promotes universities' commitment to the SDGs and recognises that universities, through teaching and research, must prepare students and researchers to find sustainable solutions to social, economic, environmental, and technical world challenges.

FIGURE 12.1 Sustainable Development Goals (SDGs). Shutterstock© 2022.

Universities around the world have been aligning educational models, graduate attributes, and programme offerings to the SDGs. They are also ensuring that curricula include real-world problem-solving projects aligned with industry and community organisations (Desha, Rowe, & Hargreaves, 2019). Some examples of these projects, which have been suggested by industry, staff, and Engineering Without Borders (EWB), include swarm robots for precision agriculture, autonomous road vehicles for high-speed motorway travel, alternative methods for electricity provision, and smart greenhouses (Murray & Horn, 2020).

In addition, specific sustainable goals have been incorporated into the engineering curriculum (Ramirez-Mendoza et al., 2020), in the form of competencies, learning activities, practical projects, and learning outcomes. For example, in response to SDG-6 (clean water and sanitation), courses from Environmental Engineering programmes provide students with the theory, processes, and design principles used for the treatment and management of water and wastewater (Shahidul, 2020).

According to Ramirez-Mendoza et al. (2020), universities themselves must implement strategic plans to address SDGs within the institution. An example of this includes the SDG Impact Report 2021 from Griffith University in Queensland, Australia (www.griffith.edu.au/sustainability), which provides information on the University's commitment to increasing the health and wellbeing of staff and students; achieving gender equality in the workplace; reducing inequalities through teaching and research; reducing e-waste; supporting sustainable resource use and production; and generating their own renewable energy through solar panels and wind turbine installations. As well as generating their own renewable energy, Griffith University has also built innovative buildings, known as "living laboratories", in which real-time information about the building's energy, water, and structural performance is provided by several sensor arrays.

The provision of clear policies and consistent practices for integrating sustainability on-campus provides an optimal environment for building students' capacity to work towards the SDGs in their future careers. In addition, through delivering impactful knowledge, and adopting SDGs in their engineering programmes, universities can connect higher education with industry, healthcare, and community partners, in the pursuit of a more equitable society and, ultimately, a better world (Purcell et al., 2019). Such synergies are necessary in order to achieve the SDGS by 2030. An important immediate target for universities to achieve these goals is to reduce the gap between engineering education and the real-world demands on engineers (Shanmuganathan, 2018). This can be achieved by adopting systematic pedagogical approaches, such as active learning.

Active learning can provide students with problem-solving competencies for sustainability-related issues. As suggested by Alm et al. (2021), one pathway for students to develop knowledge about the SDGs, and a change-agent identity towards the challenges established by the SDGs, can be divided into three stages (Figure 12.2):

1. Creating learning settings: in this stage, active learning approaches are implemented in the classroom for the understanding of sustainable development and the SDGs.

FIGURE 12.2.   Student pathway based on SDGs.

2. Learning outcomes: it is expected that by the end of the teaching term, students will have developed SDG awareness, as well as key sustainability competencies.
3. Change-agent for SDGs: in this stage, students become advocates of change, ready to promote sustainable development in their future professional careers (Larsson & Holmberg, 2018).

The question that arises from the above is: How can active learning, the starting point of this transition process, be effectively implemented in our teaching practices? Lecture-centric courses (traditional teaching) were already declining in favour of more innovative teaching practices; COVID-19 simply accelerated this transition in 2020, when educators around the world were forced to move their on-campus teaching online. This sudden shift has irrevocably changed the higher education landscape and, as stated in a *New York Times* article by Frank (2020), "online lectures are here to stay". It is proven that active learning is more effective than traditional teaching (Freeman et al., 2014), as it improves student performance, engagement, and success; so, as the lockdown restrictions eased by the end of 2020 and throughout 2021, universities around the world saw this as a motivation to implement hybrid educational models using active learning for synchronous and asynchronous teaching delivery. Envisioned as a learning method for the future of engineering education, it is important to implement it effectively irrespective of the type of teaching delivery (in-person, hybrid, or online), and either synchronously or asynchronously.

According to Kosslyn (2021), in active learning the instructor is required to design activities aiming to engage students in material that will allow them to achieve the required learning outcomes. Some learning activities commonly introduced into the classroom include case studies, problem-solving activities, discussions, and group projects. Kosslyn (2021) provides instructors with several strategies and examples of activities aiming to make online classes more effective, engaging, stimulating, and enjoyable for students. It is important to also note that existing evidence supports the efficacy of integrating active learning methodologies with blended learning strategies, where online and F2F learning experiences are combined (Baepler et al., 2014).

INNOVATION IN THE VIRTUAL CLASSROOM

The disruption to higher education in 2020 due to COVID-19 caused a drastic digital transformation, which resulted in an accelerated adoption of different teaching strategies for online environments aiming to improve student engagement and performance.

Videoconferencing platforms (Marshall & Ward, 2020; Thakker et al., 2021) such as Microsoft Teams® (Wea & Dua Kuki, 2021), Zoom® (Johnson, 2022), Google Meet® (Al-Maroof et al., 2020), GoToMeeting®, and Webex®, have become common resources for universities, as they allow educators to set virtual spaces used for teaching, meetings, webinars, training, and even social gatherings. The ease of lockdown restrictions allowed moving from fully online to hybrid delivery.

Innovative and contemporary pedagogical strategies have been adopted by educators to mitigate the challenges experienced by COVID-19, and can be used for both online and hybrid delivery modes either synchronously or asynchronously. Selected strategies that have gained considerable interest in engineering education include:

- Challenge-based learning (Membrillo-Hernández et al., 2021).
- Project-based learning (Gratchev et al., 2018; Kokotsaki et al., 2016).
- Problem-based learning (Masitoh & Fitriyani, 2018).
- Hyflex and BlendFlex (Beatty, 2019; Furse & Ziegenfuss, 2021; Miller et al., 2021).
- Flipped classroom (Furse & Ziegenfuss, 2020; Hew et al., 2020; Schwerin et al., 2021).
- Experiential learning (Ernst, 2013; Espinosa et al., 2020).
- Competency-based learning (Artner & Mecklenbrauker, 2020).

Several initiatives and resources have been developed in recent years to support the previous pedagogical strategies, which include:

- Gamification (Milosz & Milosz, 2020; Ortiz-Rojas et al., 2019).
- Mobile learning applications (Tan, 2021).
- Virtual reality (Pendergast et al., 2022).
- Artificial Intelligence (Johri, 2020; Zawacki-Richter et al., 2019).
- Worked example videos (Dart et al., 2020).
- Interactive computational tools (Espinosa & Sevgi, 2021a).
- Massive open online courses (MOOCs) (Feitosa de Moura et al., 2021).
- Remote and virtual labs (Glassey & Magalhães, 2020; Grodotzki et al., 2018).
- Three-dimensional virtual environments (Mora-Beltrán et al., 2020)

More technological innovations and teaching practices can be found in Hernandez-de-Menendez and Morales-Menendez (2019), where a comprehensive list of tools and resources selected from top QS ranking universities is presented. Some of these tools include gaming, virtual environments, audience response systems, digital assessment, and e-portfolios.

The strategies discussed above offer different approaches to education, as well as different implementation processes and required resources. Regardless of the approach adopted by educators, it is important that these strategies encourage active student participation, which improves students' skills, engagement, learning outcomes, and knowledge retention. It is also very important to provide engineering students with sufficient technical abilities and practical skills that are required in industry. In the end, the main goal of engineering education is to develop engineers who will become future leaders and innovators.

While the implementation of innovative pedagogical strategies can improve student performance, their efficacy is influenced by the instructor's competence. Nowadays, expertise in the course content is not enough to deliver successful hybrid or fully online courses. Skills in the design and implementation of blended learning strategies, as well as skills in the use of technological resources, are now essential for educators. Fortunately, some non-profit organisations, such as "Teachonline.ca" (https://teachonline.ca/), offer several resources on new technologies and developments in online and distance learning. They provide educators with teaching resources and strategies to assist them with issues such as pedagogical approaches, student support and retention, educational trends, MOOCs, access to relevant articles, webinars, training opportunities, and lists of relevant conferences.

Other organisations, such as the Institute for the Future of Education (IFE) (https://observatory.tec.mx/), provide teachers, non-academic staff, and students with free educational resources, practices, and articles. A section on post-COVID-19 resources is also available in the IFE website, which includes resources from sections related to teaching technology, shift to online teaching, mental health and wellbeing, and work-from-home (WFH) tools, courses, etc.

Online learning requires careful instructional design and planning; poor design and implementation substantially detract from the quality of the instruction in online courses (Hodges et al., 2020). Although online learning seems to be a vital alternative for the future of engineering education, challenges still persist in the access to infrastructure and resources that supports its delivery (Lockee, 2021). In a study on educational challenges in Ghana post-COVID-19, Adarkwah (2021) identified several barriers to online learning, including inadequate access to technology (i.e., personal computers), limited Internet connectivity at home (i.e., poor Internet connection and low speed), lack of constant supply of electricity, and scepticism from educators. In addition, the study provided novel strategies for educators and policymakers to ensure effective online learning, taking into consideration the needs of urban, urban poor, and rural students.

One may think that the use of blended learning strategies will eventually entirely replace traditional teaching; this is still debatable, and it is by no means the intention of this chapter to suggest the complete dismissal of traditional lectures in an online environment. Instead, and as suggested by Kosslyn (2021), the recommendation is for educators to ask: "When and how should a lecture be used?" Kosslyn (2021) discussed several strengths and weaknesses of lectures, which can be applied equally to online and traditional settings. The main takeaway is that, in a well-designed lecture, there should be a balance between active learning activities, reflective learning experiences, and short lectures.

Regardless of how courses are delivered (online or in-person), it is very important that engineering educators retain the fundamental features of engineering courses. As shown in Figure 12.3, engineering courses should be designed around three constitutive and inter-related pillars: theoretical background, computational modelling, and active experimentation (Espinosa & Sevgi, 2021b). Developing mathematical skills and the conceptual understanding of the course content are essential. Computational modelling and simulation reinforce students' understanding of the theory. Developing practical skills through problem-solving and hands-on practice are attributes highly required by industry.

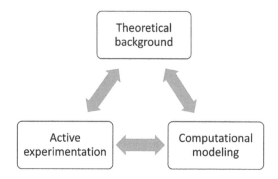

FIGURE 12.3.   Teaching and learning pillars in engineering education.

CONCLUSIONS

The higher education sector is experiencing a great transformation, and more change will occur over the next few years. Examples in this chapter demonstrate that solutions for many of the challenges associated with incorporating online learning in engineering have already been developed. This state of play offers optimism that remaining challenges will also be similarly addressed in the near future.

It is clear that online teaching is shaping the future of engineering education. As such, it is important that educators consider adopting some of the strategies discussed in this chapter, as well as those described comprehensively in the literature. Current generations of engineering students are digital natives. As discussed by Seemiller and Grace (2016), generation Z students, those born during or after 1995, are driven by different learning styles compared to previous generations, due to their varying skill sets developed as digital natives. They are active learners who can easily adopt technology for learning, as it has always been part of their daily lives. Therefore, educators must become technologically literate, and ensure that any novel practice adopted has a clear focus on learning outcomes aligned with curriculum, pedagogy, research, innovation, and industry. According to Teachonline.ca, learners can achieve their learning outcomes through a combination of self-directed, instructor-delivered, in-class and online learning: this is why we, as educators, need to be prepared and ready to support our future generation of engineers.

NOTES

1   iLecture is a system that allows access to recordings of unit lectures or other video-based resources.
2   Blackboard is a virtual learning environment and learning management system developed by Blackboard Inc. (https://www.blackboard.com/). It is a common learning management system used at several universities around the world.

REFERENCES

Adarkwah, M. A. (2021). "I'm not against online teaching, but what about us?": ICT in Ghana post-Covid-19. *Education and Information Technologies, 26*(2), 1665–1685. doi:10.1007/s10639-020-10331-z

Alm, K., Melén, M., & Aggestam-Pontoppidan, C. (2021). Advancing SDG competencies in higher education: Exploring an interdisciplinary pedagogical approach. *International Journal of Sustainability in Higher Education, 22*(6), 1450–1466. doi:10.1108/IJSHE-10-2020-0417

Al-Maroof, R. S., Salloum, S. A., Hassanien, A. E., & Shaalan, K. (2020). Fear from COVID-19 and technology adoption: The impact of Google Meet during Coronavirus pandemic. *Interactive Learning Environments*, 1–16. doi:10.1080/10494820.2020.1830121

Alves, A. C., Fernandes, S., & Uebe-Mansur, A. F. (2021). *Online assessment: More student cheating than on-site?* Paper presented at the International Symposium on Project Approaches in Engineering Education/Active Learning in Engineering Education Workshop/International Conference on Active Learning in Engineering Education., Braga, Portugal.

Artner, G., & Mecklenbrauker, C. F. (2020). The competence-oriented teaching of antennas, propagation, and wireless communications: Enabling independent students. *IEEE Antennas and Propagation Magazine, 62*(2), 43–49. doi:10.1109/MAP.2020.2969250

Asgari, S., Trajkovic, J., Rahmani, M., Zhang, W., Lo, R. C., & Sciortino, A. (2021). An observational study of engineering online education during the COVID-19 pandemic. *PLOS ONE, 16*(4). doi:10.1371/journal.pone.0250041

Attallah, B. (2020). *Post-COVID-19 higher education empowered by virtual worlds and applications*. Paper presented at the 7th International Conference on Information Technology Trends, United Arab Emirates.

Baepler, P., Walker, J. D., & Driessen, M. (2014). It's not about seat time: Blending, flipping, and efficiency in active learning classrooms. *Computers & Education, 78*, 227–236. doi:10.1016/j.compedu.2014.06.006

Bawa, P. (2016). Retention in online courses: Exploring issues and solutions—A literature review. *SAGE Open, 6*(1). doi:10.1177/2158244015621777

Beatty, B. J. (2019). *Hybrid-flexible course design: Implementing student-directed hybrid classes*. EdTechBooks.org

Biggs, A., & Espinosa, H. G. (2021). *Intervention to enhance PsyCap in EM courses*. Paper presented at the *2021 IEEE International Symposium on Antennas and Propagation and USNC-URSI Radio Science Meeting (APS/URSI)*.

Bilen, E., & Matros, A. (2021). Online cheating amid COVID-19. *Journal of Economic Behavior & Organization, 182*, 196–211. doi:10.1016/j.jebo.2020.12.004

Boud, D., & Molloy, E. (2013). Rethinking models of feedback for learning: The challenge of design. *Assessment & Evaluation in Higher Education, 38*(6), 698–712. doi:10.1080/02602938.2012.691462

Brasco, C., Chen, L., Hojnacki, M., & Krishnan, C. (2019). Transformation 101: How universities can overcome financial headwinds to focus on their mission. Retrieved from https://www.mckinsey.com/business-functions/transformation/our-insights/transformation-101-how-universities-can-overcome-financial-headwinds-to-focus-on-their-mission

Brookfield, S. D. (2006). *The Skillful Teacher: On Technique, Trust, and Responsiveness in the Classroom*. (2nd ed.).

Cavalcanti, A. P., Barbosa, A., Carvalho, R., Freitas, F., Tsai, Y.-S., Gašević, D., & Mello, R. F. (2021). Automatic feedback in online learning environments: A systematic literature review. *Computers and Education: Artificial Intelligence, 2*, 100027. doi:10.1016/j.caeai.2021.100027

Clark, R. M., Kaw, A. K., & Braga Gomes, R. (2022). Adaptive learning: Helpful to the flipped classroom in the online environment of COVID? *Computer Applications in Engineering Education, 30*(2), 517–531. doi:10.1002/cae.22470

Council of Australasian University Leaders in Learning and Teaching. (2022). Contemporary Approaches to University Teaching MOOC. Retrieved from https://www.caullt.edu.au/announcements/contemporary-approaches-to-university-teaching-mooc/

Cruz, M. L., van den Bogaard, M. E., Saunders-Smits, G. N., & Groen, P. (2020). Testing the validity and reliability of an instrument measuring engineering students' perceptions of transversal competency levels. *IEEE Transactions on Education, 64*(2), 180–186.

Damary, R., Markova, T., & Pryadilina, N. (2017). Key challenges of online education in multi-cultural context. *Procedia-Social and Behavioral Sciences*, *237*, 83–89.

Darby, F., & Lang, J. M. (2019). *Small Teaching Online: Applying Learning Science in Online Classes*. Jossey-Bass; Wiley.

Dart, S., Pickering, E., & Dawes, L. (2020). Worked example videos for blended learning in undergraduate engineering. *Advances in Engineering Education*, *8*(2), 1–22.

Desha, C., Rowe, D., & Hargreaves, D. (2019). A review of progress and opportunities to foster development of sustainability-related competencies in engineering education. *Australasian Journal of Engineering Education*, *24*(2), 61–73. doi:10.1080/22054952. 2019.1696652

Dias, A., Scavarda, A., Silveira, H., Scavarda, L. F., & Kondamareddy, K. K. (2021). The online education system: COVID-19 demands, trends, implications, challenges, lessons, insights, opportunities, outlooks, and directions in the work from home. *Sustainability*, *13*(21). doi:doi:10.3390/su132112197

Ding, F., & Riccucci, N. M. (2020). The value of alumni networks in responding to the public administration theory and practice: Evidence from the COVID-19 pandemic in China. *Administrative Theory and Praxis*, *42*(4), 588–603. doi:10.1080/10841806.2020. 1798694

Dow, M. J. (2008). Implications of social presence for online learning: A case study of MLS students. *Journal of Education for Library and Information Science*, *49*(4), 231–242.

Ebert, K., Axelsson, L., & Harbor, J. (2015). Opportunities and challenges for building alumni networks in Sweden: A case study of Stockholm University. *Journal of Higher Education Policy and Management*, *37*(2), 252–262. doi:10.1080/1360080X.2015.1019117

Ernst, J. V. (2013). Impact of experiential learning on cognitive outcome in technology and engineering teacher preparation. *Journal of Technology Education*, *24*(2), 31–40.

Espinosa, H. G. & Sevgi, L. (2021a). Interactive computational tools for electromagnetics education enhancement. In K. T. Selvan & K. F. Warnick (Eds.), *Teaching Electromagnetics: Innovative Approaches and Pedagogical Strategies* (pp. 77–102). CRC Press.

Espinosa, H. G., & Sevgi, L. (2021b). *Effective electromagnetics teaching, no matter what!* Paper presented at the *2021 IEEE International Symposium on Antennas and Propagation and USNC-URSI Radio Science Meeting (APS/URSI)*.

Espinosa, H. G., Fickenscher, T., Littman, N., & Thiel, D. V. (2020). *Teaching wireless communications courses: An experiential learning approach*. Paper presented at the *2020 14th European Conference on Antennas and Propagation (EuCAP)*.

Espinosa, H. G., Khankhoje, U., Furse, C., Sevgi, L., & Rodriguez, B. S. (2021). Learning and teaching in a time of pandemic. In K. T. Selvan & K. F. Warnick (Eds.), *Teaching Electromagnetics: Innovative Approaches and Pedagogical Strategies* (pp. 219–237). CRC Press.

Feitosa de Moura, V., Alexandre de Souza, C., & Noronha Viana, A. B. (2021). The use of Massive Open Online Courses (MOOCs) in blended learning courses and the functional value perceived by students. *Computers & Education*, *161*, 104077. doi:10.1016/j. compedu.2020.104077

Fluck, A. E. (2019). An international review of e-Exam technologies and impact. *Computers & Education*, *132*, 1–15. doi:10.1016/j.compedu.2018.12.008

Frank, R. H. (2020). Don't kid yourself: Online lectures are here to stay. *New York Times*. Retrieved from https://www.nytimes.com/2020/06/05/business/online-learning-winner-coronavirus.html

Freeman, S., Eddy, S. L., McDonough, M., Smith, M. K., Okoroafor, N., Jordt, H., & Wenderoth, M. P. (2014). Active learning increases student performance in science, engineering, and mathematics. *Proceedings of the National Academic of Sciences*, *111*(23), 8410–8415. doi:10.1073/pnas.1319030111

Froyd, J. E., Lord, S. M., Ohland, M. W., Prahallad, K., Lindsay, E. D., & Dicht, B. (2014, 22–25 October). *Scenario planning to envision potential futures for engineering education.* Paper presented at the *2014 IEEE Frontiers in Education Conference (FIE) Proceedings.*

Furse, C., & Ziegenfuss, D. (2020). A busy professor's guide to sanely flipping your classroom: Bringing active learning to your teaching practice. *IEEE Antennas and Propagation Magazine, 62*(2), 31–42. doi:10.1109/MAP.2020.2969241

Furse, C., & Ziegenfuss, D. (2021). HyFlex Flipping: Combining in-person and online teaching for the flexible generation. In K. T. Selvan & K. F. Warnick (Eds.), *Teaching Electromagnetics: Innovative Approaches and Pedagogical Strategies* (pp. 201–218). CRC Press.

Geisinger, B. N., & Raman, D. R. (2013). Why they leave: Understanding student attrition from engineering majors. *International Journal of Engineering Education, 29*, 914–925.

Glassey, J., & Magalhães, F. D. (2020). Virtual labs – Love them or hate them, they are likely to be used more in the future. *Education for Chemical Engineers, 33*, 76–77. doi:10.1016/j.ece.2020.07.005

Gottardello, D., & Karabag, S. F. (2022). Ideal and actual roles of university professors in academic integrity management: A comparative study. *Studies in Higher Education, 47*(3), 526–544. doi:10.1080/03075079.2020.1767051

Gratchev, I., Jeng, D. S., & Oh, E. (2018). *Soil Mechanics through Project-Based Learning.* CRC Press.

Grodotzki, J., Ortelt, T. R., & Tekkaya, A. E. (2018). Remote and virtual labs for engineering Education 4.0: Achievements of the ELLI project at the TU Dortmund University. *Procedia Manufacturing, 26*, 1349–1360. doi:10.1016/j.promfg.2018.07.126

Grodotzki, J., Upadhya, S., & Tekkaya, A. E. (2021). Engineering education amid a global pandemic. *Advances in Industrial and Manufacturing Engineering, 3*, 100058. doi:10.1016/j.aime.2021.100058

Hart, J. (2021). Top 300 tools for learning 2021. Retrieved from https://www.toptools4learning.com/

Hartman, V. F. (1995). Technical and learning style preferences: Transitions through technology. *VCCA Journal, 9*, 18–20.

Henry, J. V., & Oliver, M. (2022). Who will watch the watchmen? The ethico-political arrangements of algorithmic proctoring for academic integrity. *Postdigital Science and Education, 4*(2), 330–353. doi:10.1007/s42438-021-00273-1

Hernandez-de-Menendez, M., & Morales-Menendez, R. (2019). Technological innovations and practices in engineering education: A review. *International Journal on Interactive Design and Manufacturing (IJIDeM), 13*(2), 713–728. doi:10.1007/s12008-019-00550-1

Hew, K. F., Jia, C., Gonda, D. E., & Bai, S. (2020). Transitioning to the "new normal" of learning in unpredictable times: Pedagogical practices and learning performance in fully online flipped classrooms. *International Journal of Educational Technology in Higher Education, 17*(1), 57. doi:10.1186/s41239-020-00234-x

Hill, C., & Lawton, W. (2018). Universities, the digital divide and global inequality. *Journal of Higher Education Policy and Management, 40*(6), 598–610. doi:10.1080/1360080X.2018.1531211

Hodges, C., Moore, S., Lockee, B., Trust, T., & Bond, A. (2020). The difference between emergency remote teaching and online learning. *Educause Review, 27*, 1–12.

IAOE/GOLC. (2022). GOLC – Online Lab Award. Retrieved from http://online-engineering.org/GOLC_online-lab-award.php

Iyer, A. K., Smyth, B. P., Semple, M., & Barker, C. (2021). Going remote: Teaching microwave engineering in the age of the global pandemic and beyond. *IEEE Microwave Magazine, 22*(11), 64–77. doi:10.1109/MMM.2021.3102649

Janke, S., Rudert, S. C., Petersen, Ä., Fritz, T. M., & Daumiller, M. (2021). Cheating in the wake of COVID-19: How dangerous is ad-hoc online testing for academic integrity? *Computers and Education Open*, *2*, 100055. doi:10.1016/j.caeo.2021.100055

Johnson, A. (2022). *Online Teaching with ZOOM: A Guide for Teaching and Learning with Videoconference Platforms*. Aaron Johnson, USA.

Johri, A. (2020). Artificial intelligence and engineering education. *Journal of Engineering Education*, *3*, 358–361.

Khan, Z. H., & Abid, M. I. (2021). Distance learning in engineering education: Challenges and opportunities during COVID-19 pandemic crisis in Pakistan. *The International Journal of Electrical Engineering & Education*. doi:10.1177/0020720920988493

Khoo, E. G. L., Peter, M., Scott, J. B., Round, W. H., & Cowie, B. (2018). *How we flipped an engineering course*. Paper presented at the *Centre for Learning and Research in Higher Education*. University of Auckland, Auckland, New Zealand. https://hdl.handle.net/10289/13026

Kochmar, E., Vu, D. D., Belfer, R., Gupta, V., Serban, I. V., & Pineau, J. (2020). Automated personalized feedback improves learning gains in an intelligent tutoring system. In I. I. Bittencourt, M. Cukurova, K. Muldner, R. Luckin, & E. Millán (Eds.), *Artificial Intelligence in Education* (pp. 140–146). Cham: Springer.

Kokotsaki, D., Menzies, V., & Wiggins, A. (2016). Project-based learning: A review of the literature. *Improving Schools*, *19*(3), 267–277. doi:10.1177/1365480216659733

Kosslyn, S. M. (2021). *Active Learning Online: Five Principles that make Online Courses Come Alive*. Alinea Learning.

Krambia Kapardis, M., & Spanoudis, G. (2022). Lessons learned during Covid-19 concerning cheating in e-examinations by university students. *Journal of Financial Crime*, *29*(2), 506–518. doi:10.1108/JFC-05-2021-0105

Kuley, E. A., Maw, S., & Fonstad, T. (2015). *Engineering student retention and attrition literature review*. Paper presented at the *2015 Canadian Engineering Education Association Conference*, Ontario.

Larsson, J., & Holmberg, J. (2018). Learning while creating value for sustainability transitions: The case of Challenge Lab at Chalmers University of Technology. *Journal of Cleaner Production*, *172*(4411–4420). doi:10.1016/j.jclepro.2017.03.072

Lockee, B. B. (2021). Online education in the post-COVID era. *Nature Electronics*, *4*(1), 5–6. doi:10.1038/s41928-020-00534-0

Marshall, D., & Ward, L. (2020). Let's collaborate! Technology, literacy, and teaching during a pandemic. *Technology and Engineering Teacher*, *80*(1), 30–31.

Masitoh, L. F., & Fitriyani, H. (2018). Improving students' mathematics self-efficacy through problem-based learning. *Malikussaleh Journal of Mathematics Learning (MJML)*, *1*(1), 26–30. doi:10.29103/mjml.v1i1.679

Means, B., Neisler, J., & Langer Research Associates. (2020). *Suddenly Online: A National Survey of Undergraduates during the COVID-19 Pandemic*. San Mateo, CA.

Membrillo-Hernández, J., de Jesús Ramírez-Cadena, M., Ramírez-Medrano, A., García-Castelán, R. M. G., & García-García, R. (2021). Implementation of the challenge-based learning approach in Academic Engineering Programs. *International Journal on Interactive Design and Manufacturing (IJIDeM)*, *15*(2), 287–298. doi:10.1007/s12008-021-00755-3

Miller, A. N., Sellnow, D. D., & Strawser, M. G. (2021). Pandemic pedagogy challenges and opportunities: Instruction communication in remote, HyFlex, and BlendFlex courses. *Communication Education*, *70*(2), 202–204. doi:10.1080/03634523.2020.1857418

Milosz, M., & Milosz, E. (2020). *Gamification in engineering education – A preliminary literature review*. Paper presented at the *2020 IEEE Global Engineering Education Conference (EDUCON)*.

Moore, K., Jones, C., & Frazier, R. S. (2017). Engineering education for Generation Z. *American Journal of Engineering Education*, *8*(2), 111–126. doi:10.19030/ajee.v8i2.10067

Mora-Beltrán, C. E., Rojas, A. E., & Mejía-, C. (2020). An immersive experience in the virtual 3D Virbela environment for ... development in undergraduate students during the COVID-19 quarantine. Paper presented at the ICAIW 2020: Workshops at the Third International Conference on Applied Informatics, Ota, Nigeria.

Murray, P. B., & Horn, R. (2020). Interdisciplinary ... In P. Kapranos (Ed.), The Interdisciplinary Future of Engineering Education... Teaching and Learning (pp. 55–76). Routledge. ...king through Boundaries in...

Nguyen, J. G., Keuseman, K. J., & Humston, J. J. (2020). Min... online cheating for online assessments during COVID-19 pandemic. Journal of Chemical Education, 97(9), 3429–3435. doi:10.1021/acs.jchemed.0c00790

Ortiz-Rojas, M., Chiluiza, K., & Valcke, M. (2019). Gamification... through leaderboards: An empirical study in engineering education. Computer Applications in Engineering Education, 27(4), 777–788. doi:10.1002/cae.12116

Park, J. J., Park, M., Jackson, K., & Vanhoy, G. (2020). Remote engineering education under COVID-19 pandemic environment. International Journal of Multidisciplinary Perspectives in Higher Education, 5(1), 160–166.

Peimani, N., & Kamalipour, H. (2021). Online education in the post-COVID-19 era: Students' perception and learning experience. Education Sciences, 11(10), 633. doi:10.3390/educsci11100633

Pendergast, D., O'Brien, M., Prestridge, S., & Exley, B. (2022). Self-efficacy in a 3-dimensional virtual reality classroom – Initial teacher education students' experiences. Education Sciences, 12(6), 368. doi:10.3390/educsci12060368

Purcell, W. M., Henriksen, H., & Spengler, J. D. (2019). Universities as the engine of transformational sustainability toward delivering the sustainable development goals. International Journal of Sustainability in Higher Education, 20(8), 1343–1357. doi:10.1108/IJSHE-02-2019-0103

Qadir, J., & Al-Fuqaha, A. (2020). A student primer on how to thrive in engineering education during and beyond COVID-19. Education Sciences, 10(9), 236.

Qamar, S. Z., Pervez, T., & Al-Kindi, M. (2019, November). Engineering education: Challenges, opportunities, and future trends. Paper presented at the Proceedings of the International Conference on Industrial Engineering and Operations Management, Saudi Arabia.

Rahman, A., & Ilic, V. (2018). Blended Learning in Engineering Education: Recent Developments in Curriculum, Assessment and Practice. CRC Press.

Raje, S., & Stitzel, S. (2020). Strategies for effective assessments while ensuring academic integrity in general chemistry courses during COVID-19. Journal of Chemical Education, 97(9), 3436–3440. doi:10.1021/acs.jchemed.0c00797

Ramirez-Mendoza, R. A., Morales-Menendez, R., Melchor-Martinez, E. M., Iqbal, H. M. N., Parra-Arroyo, L., Vargas-Martínez, A., & Parra-Saldivar, R. (2020). Incorporating the sustainable development goals in engineering education. International Journal on Interactive Design and Manufacturing (IJIDeM), 14(3), 739–745. doi:10.1007/s12008-020-00661-0

Ramo, N. L., Lin, M., Hald, E. S., & Huang-Saad, A. (2021). Synchronous vs. asynchronous vs. blended remote delivery of introduction to biomechanics course. Biomedical Engineering Education, 1(1), 61–66. doi:10.1007/s43683-020-00009-w

Rouvrais, S., Gerwel Proches, C., Andunsson, H., Chelin, N., Liem, I., & Tudela Villalonga, L. (2020, November 16). Preparing 5.0 engineering students for an unpredictable post-COVID world. Paper presented at the WEEF/GEDC 2020: World Engineering Education Forum and the Global Engineering Deans Council, Cape Town, South Africa.

Ryan, M. D., & Reid, S. A. (2016). Impact of the flipped classroom on student performance and retention: A parallel controlled study in general chemistry. Journal of Chemical Education, 93(1), 13–23. doi:10.1021/acs.jchemed.5b00717

de Sande, J. C. G. (2015). Calculated questions and e-cheating: A case study. In M. Carmo (Ed.), Education Applications and Developments (Vol. 5, pp. 92–102). Portugal: inScience Press.

Schwerin, B., Espinosa, H. G., Gra... & Lohmann, G. (2021). *Enhancing maths teaching resources: Topic video ...torial streaming development.* Paper presented at the Research in Enginee...ucation Symposium & Australasian Association for ...nce, Perth, Australia.

Engineering Education ...*Generation Z Goes to College.* Jossey-Bass; Wiley.

Seemiller, C., & Grace, M. ...oon, D. (2021). The impact of artificial intelligence on

Seo, K., Roll, S., Fels, ...tion in online learning. *International Journal of Educational* learner-instructor *Education*, 18(1), 54–76. doi:10.1186/s41239-021-00292-9 Technology in H...gineering education for achieving sustainable development goals

Shahidul, M. I. (202...g the paths for challenging climate change and Covid 19. *Science* by 2030: Re...ahore), 32(4), 403–410. Internation...

Shanmuganath...3. (2018). Engineering education online. In A. Rahman & V. Ilic (Eds.), Blende...earning in Engineering Education: Recent Developments in Curriculum, Asse...ent and Practice. CRC Press.

Smith, S. ...020). Fourteen Simple Strategies to Reduce Cheating on Online Examinations. ...etrieved from https://www.facultyfocus.com/articles/educational-assessment/fourteen-simple-strategies-to-reduce-cheating-on-online-examinations/

Snijders, I., Wijnia, L., Rikers, R. M. J. P., & Loyens, S. M. M. (2019). Alumni loyalty drivers in higher education. *Social Psychology of Education*, 22(3), 607–627. doi:10.1007/s11218-019-09488-4

Talbert, R. (2017). *Flipped Learning: A Guide for Higher Education Faculty.* Stylus Publishing, LLC.

Tan, E. L. (2021). Computational electromagnetics and mobile apps for electromagnetics education. In K. T. Selvan & K. F. Warnick (Eds.), *Teaching Electromagnetics: Innovative Approaches and Pedagogical Strategies* (pp. 103–127). CRC Press.

Tapanes, M. A., Smith, G. G., & White, J. A. (2009). Cultural diversity in online learning: A study of the perceived effects of dissonance in levels of individualism/collectivism and tolerance of ambiguity. *The Internet and Higher Education*, 12(1), 26–34.

Tarigan, R. N., Nadlifatin, R., & Subriadi, A. P. (2021). *Academic dishonesty (cheating) in online examination: A literature review.* Paper presented at the *2021 International Conference on Computer Science, Information Technology, and Electrical Engineering,* Indonesia.

Thakker, S. V., Parab, J., & Kaisare, S. (2021). Systematic research of e-learning platforms for solving challenges faced by Indian engineering students. *Asian Association of Open Universities Journal*, 16(1), 1–19. doi:10.1108/AAOUJ-09-2020-0078

Vergara, D., Fernández-Arias, P., Extremera, J., Dávila, L. P., & Rubio, M. P. (2022). Educational trends post-COVID-19 in engineering: Virtual laboratories. *Materials Today: Proceedings*, 49, 155–160. doi:10.1016/j.matpr.2021.07.494

Vesikivi, P., Lakkala, M., Holvikivi, J., & Muukkonen, H. (2020). The impact of project-based learning curriculum on first-year retention, study experiences, and knowledge work competence. *Research Papers in Education*, 35(1), 64–81. doi:10.1080/02671522.2019.1677755

Waltzer, T., & Dahl, A. (2022). Why do students cheat? Perceptions, evaluations, and motivations. *Ethics & Behavior*, 1–21. doi:10.1080/10508422.2022.2026775

Wea, K. N., & Dua Kuki, A. (2021). Students' perceptions of using Microsoft Teams application in online learning during the Covid-19 pandemic. *Journal of Physics: Conference Series*, 1842(1), 012016. doi:10.1088/1742-6596/1842/1/012016

Zawacki-Richter, O., Marín, V. I., Bond, M., & Gouverneur, F. (2019). Systematic review of research on artificial intelligence applications in higher education – Where are the educators? *International Journal of Educational Technology in Higher Education*, 16(1), 39. doi:10.1186/s41239-019-0171-0

Index